SAVOR

SAVOR

Mindful Eating, Mindful Life

Thich Nhat Hanh

and

Lilian Cheung, D.Sc., R.D.

HarperOne
An Imprint of HarperCollins*Publishers*

HarperOne

This book is written as a source of information only. The information contained in this book should by no means be considered a substitute for the advice of a qualified medical professional, who should always be consulted before beginning any new diet, exercise, or other health program.

All efforts have been made to ensure the accuracy of the information contained in this book as of the date published. The authors and the publisher expressly disclaim responsibility for any adverse effects arising from the use or application of the information contained herein.

Figure 3.1, p. 63, reprinted from *The Heart of the Buddha's Teaching: Transforming Suffering into Peace, Joy & Liberation* (1998) by Thich Nhat Hanh with permission of Parallax Press, Berkeley, California, www.parallax.org.

Table 6.1, p. 150, from U.S. Department of Health and Human Services, *2008 Physical Activity Guidelines for Americans* (Washington, DC: U.S. Department of Health and Human Services, 2008).

HarperCollins books may be purchased for educational, business, or sales promotional use. For information please write: Special Markets Department, HarperCollins Publishers, 10 East 53rd Street, New York, NY 10022.

HarperCollins website: http://www.harpercollins.com
HarperCollins®, ®, and HarperOne™ are
trademarks of HarperCollins Publishers

FIRST HARPERCOLLINS PAPERBACK EDITION PUBLISHED IN 2011

Library of Congress Cataloging-in-Publication Data is available upon request.

ISBN 978-0-06-169770-8

12 13 14 15 RRD(H) 10 9 8 7 6 5 4 3

CONTENTS

PART 3
Individual and Collective Effort

LIST OF ILLUSTRATIONS

FOREWORD

Most books about nutrition and diet stress the calorie content of foods, saturated fats, trans fats, carbohydrates, portion size, and the balance between energy intake and energy expended. This book is different. This book is not only about *what* to eat. This book also teaches *how* to eat.

If you are someone who is concerned about your weight, perhaps you have tried dieting in the past. Maybe you have gone from cutting calories, to excluding fats, to avoiding "carbs," to taking a chance on the grapefruit diet or some other fad. Perhaps you have even lost weight—only to gain it right back within a few months. You know something is wrong and you feel not in control of your own body. You are convinced that something has to change, but where do you begin?

The answer, provided in this book, is not to begin with what you put on your plate. Rather, you begin with what is already inside you, with your awareness and experience of every moment you live—what the authors call *mindfulness.* While the concept of mindfulness derives from Buddhist teaching, anyone can become more mindful in nourishing our bodies. Many distractions in daily life reinforce the mindless ingestion of food, and mindless eating is a strong driver of weight gain and obesity. With awareness and practice, it is possible to become more mindful in our eating—and in our lives. This book tells you how to do it.

Authored by an eminent spiritual leader and a renowned nutritionist, this work infuses science into wisdom and wisdom into science. It is a practical guide to eating mindfully and points the way to attain a healthier weight and a more satisfying life.

Harvey V. Fineberg, M.D., Ph.D.
President
Institute of Medicine
The National Academies
Washington, D.C.

ACKNOWLEDGMENTS

Savor: Mindful Eating, Mindful Life has come to be because the conditions are right. This book would not be possible without the hard work of scientists, humanitarians, health experts, policy makers, spiritual teachers, and practitioners across many generations, and all others known and unknown who have dedicated their lives to making our world a healthier and more compassionate place. Our deepest gratitude to them all.

We wish to thank Dr. Harvey V. Fineberg for writing the foreword and Dr. Walter Willett for reviewing our book. We would like to express our special appreciation to Sari Kalin, Hank Dart, and Joanne Levy for assisting in the research and the preparation of the manuscript, and our sincere thanks to our loving *sanghas,* friends, and families for their kind support throughout this journey. Finally, we wish to thank our editor at HarperOne, Gideon Weil, and his associates, for all their help in making this book a reality.

Thich Nhat Hanh & Lilian Cheung
December 2009

Introduction

IF YOU HAVE PICKED up *Savor* because you want more peace and happiness in everyday life, then you are in the right place. If you have picked up *Savor* because you or someone you love is struggling with weight and in need of practical solutions, then this book is for you also.

Learning to eat and live mindfully is the key to experiencing health and peace. This vision brought us, a Zen Buddhist Master and a nutritionist, together to take a fresh look at mindfulness and the problem of obesity that is spreading across the globe. It is clear that standard approaches are not working to stem the tide of weight gain and the burden it creates for people and communities. More than a billion people worldwide are overweight—so many, in fact, that scientists call it an obesity epidemic. And while this is certainly an apt description of the outcome of eating too much and not moving enough, it does not fully describe what is going on. It is also a worldwide crisis largely brought on by social trends that distract and prevent us from doing the things that keep us in balance, healthy, and connected with our inner selves and our place in the world. *Savor* is a guide to help us all reconnect with these different aspects of our lives that together can improve our weight, our well-being, and the well-being of our world.

Common sense tells us that to lose weight, we must eat less and exercise more. That is easier said than done, however. Many of us know that we need to eat less, and we need to be more active. But somehow we get stuck. We start on a weight-loss program with only good intentions, but when we cannot stay on track, we feel disappointed and discouraged. We become pessimistic, more and more unhappy with our weight. We spend hours worrying about our future, blaming ourselves for what we've eaten or how inactive we've been in the past, and completely missing the present moment—the moment in which we actually have the power to make real change in our lives.

To end our struggle, we must learn to not let regret, worry, or fear dominate our life in the present moment. Each minute we spend worrying about the future and regretting the past is a minute we miss in our appointment with life—a missed opportunity to engage life and to see that each moment gives us the chance to change for the better, to experience peace and joy. The practice of being fully present in each moment is called *mindfulness.* It is an ancient Buddhist approach to living that helps us to be in the here and now—and to end our struggle with weight.

To be mindful is to be completely aware of what is happening in the present, to be fully aware of all that is going on within ourselves and all that is happening around us, from moment to moment, without judgment or preconceived notions. Although mindfulness has been taught in Eastern meditation trainings, it is not some mystical or esoteric practice that is hard to learn. People in all walks of life have followed this set of age-old practices on their path to health, well-being, peace, and happiness for over twenty-five hundred years.

When we walk and our mind is aware of each step that we plant firmly on the ground, we are already practicing mindfulness. When we eat and our mind is aware of each bite, savoring the taste and the nourishment it gives us, we are already practicing mindfulness. To be mindful of something, we need to learn to be fully present for an instant and look deeply into that something. We must first stop our wandering mind in order to begin to engage it in what is there in the present moment. When we are mindful of what we are doing, we learn to keep our appointment with life. This awareness of the present moment gives us the opportunity and the tools to touch peace and joy, to see the true nature of who we are and how we are related to everything else, and to end our struggle with weight.

In *Savor* we share with you how to live mindfully, and in particular how living mindfully can help you take care of your weight problem in a peaceful and sustainable way. Throughout the book, we show you how to easily adopt the practice of mindfulness and integrate it into eating, physical activity, and all other facets of your daily life so that mindfulness gradually becomes a core part of your being.

Dealing with our overweight—or with any of our life's difficulties, for that matter—is not a battle to be fought. Instead, we must learn how to make friends with our hardships and challenges. They are there to help us; they are natural opportunities for deeper understanding and transformation, bringing us more joy and peace as we learn to work with them. With mindfulness practice, we gain insight into the roots of our difficulties.

You will learn to observe and be more aware of your motivations for and obstacles to staying healthy. Why do you eat what you eat? How do you eat, and how do you feel after you eat? What attitudes do you have toward physical activity? What are the barriers—physical, psychological, cultural, and environmental—that prevent you from eating well and staying active? As you become more aware of your body and of the feelings, thoughts, and realities that prevent you from taking health-enhancing actions, you will realize what you need to do individually and what types of community and social support you need in order to change your behavior. With these insights, you will be better equipped to break through your barriers to a healthy lifestyle and weight control. You can start to make changes step-by-step, establishing healthier habits and instilling more and more peace within yourself.

With the hectic pace of modern living, we recognize that it is very challenging for people to add more tasks to their "to do" list. As you will see, mindfulness practice does not need to be another "add-on." Its beauty lies in the fact that it can be fully and easily integrated into every act of our daily living, reminding us to live the present moment fully.

With mindfulness, we can choose *how* to live our lives now. We can seize any moment and begin anew. It is as simple as taking a few in-breaths and out-breaths here and there throughout the day—while answering e-mails, waiting in line, or sitting in traffic. It is absolutely within anyone's reach. Take small steps every day, and be persistent. Small steps will add up over time. With the constant practice of mindfulness, you will feel more alive and in the flow of life. You will find more stillness amid the chaos. You will have a better understanding of yourself and all that is around you. The fog that veils your well-being will gradually dissipate, allowing you to touch the joy and peace that have always been inside you. The Buddha did not foresee many of the modern-day problems that we face, including the obesity epidemic, but his teachings are timeless. They are foundations for understanding and processes for gaining insight. As people have found over many generations, solidity, freedom, calm, and joy are the fruits of mindful living.

By combining the sage Buddhist philosophy with the science of nutrition, we can contribute to a better understanding of our bodies and minds. We invite you to embark on this mindfulness journey, just as others have over many generations, to help you end your struggle with weight and improve your health, the health of those around you, and the well-being of the world in which we live.

About This Book

The first part of *Savor* (chapters 1 through 4) provides a Buddhist perspective on weight control and well-being. We begin the book by offering you a new way of looking at your weight problem—through the lens of the Four Noble Truths, the most fundamental Buddhist teaching. You first need to acknowledge that you have a weight problem and reflect on whether or not you are ready to make changes and are committed to making them. If you are, you need to know what has led to your being overweight. Through a series of questions for self-inquiry, you will gain a better understanding of the reasons that you are overweight. You will realize that you do have the power to stop the unhealthy, mindless routines that led you to become overweight, if you so choose. And you will realize that mindfulness practice is an effective path toward a healthier way of living and a healthier weight.

Next, we offer you a taste of mindfulness—a meditation on eating an apple. With mindfulness, the simple act of eating an apple becomes a profound experience. It opens up our awareness that the apple is a manifestation of our world and that it cannot come into being in isolation. The apple is dependent on everything else for its existence, reminding us that we, too, are constantly supported by the effort of many beings so that we can enjoy the apple.

We then invite you to look at your personal makeup. We explore how everything you see, taste, smell, hear, touch, and think affects your weight and all other aspects of your daily life. Diet books typically focus only on how foods immediately benefit the body and rarely address the psychosocial, cultural, and environmental factors that affect our habits of eating and physical activity. It is clear that we are not only what we eat; we are what we consume through all our senses. What we eat and how we eat are influenced by our ancestors, parents, and culture, the food businesses, the media, and societal forces. To maintain a healthy weight and lifestyle, we must look carefully at the interrelationships of our body, our mind, and society at large.

As you begin to understand your difficulty with weight, an important step is to learn to empower yourself to transform this difficulty. The Buddha's Discourse on the Four Foundations of Mindfulness elaborates on exercises that can lead to transformation and healing. We take you through the salient points of the four foundations of mindfulness practice for effective transformation, showing how to instill joy and peace in yourself as well as overcome your destructive habits. We describe a process and practice that will enhance your awareness of your body, feelings, mind, and all perceptions, and how they are intimately related to changing your eating and physical activity habits. You will learn basic breathing techniques to calm your body, feelings, and mind, and you will learn how to see yourself in relation to everything that is around you. With continuous practice, you will build up the confidence to recognize the nature and cause of your suffering and then transform it at its base, to uproot long-held negative beliefs, habits, and feelings.

The second part of *Savor* (chapters 5 through 7) brings mindfulness practice to your daily life. All of us eat and drink numerous times a day. These are wonderful opportunities for us to practice mindfulness. Paying attention to what we eat and drink as well as to how we eat and drink brings us nourishment, sustaining not only our own body and mind but also the well-being of our world for future generations. We provide you with a scientific primer on healthy eating and drinking, to help you decide what to eat and drink, and to offer concrete ways for you to eat and drink more mindfully. Through a series of questions, you begin to establish your own mindful-eating goals and personalized mindful-eating plan so that you can savor your foods and drinks while maintaining health and well-being.

We move our bodies every day, though our twenty-first-century lifestyle has rendered us much more sedentary. Moving mindfully not only helps us burn more calories and maintain our health but also is a valuable opportunity to practice mindfulness in action. In chapter 6, we offer mindful ways to help you maintain the level of physical activity that you need to improve your weight and health. Included are reflections to help you create your personalized mindful-moving plan.

Mindful eating and moving are both held in a larger net of general everyday, moment-to-moment mindfulness. To enhance your rhythm of mindful living, we propose simple tools to increase your practice of mindfulness throughout your waking hours. By creating a strategic mindfulness living plan, you can turn even the most mundane chore into an opportunity to practice mindfulness and nourish yourself.

As you steadily progress in your journey toward a healthy weight, you will realize that your ability to make wholesome choices and follow a healthier lifestyle depends not only on you, but also on all that is around you. We are all connected and interdependent. What one does will affect everything and everyone else, and what everything and everyone else does will have an effect on each individual. Mindful living and weight control are not just an individual matter. We must also take collective action to make our environment less toxic and more supportive of healthy eating and active living. In the third part of the book (chapter 8), we explore ways we can help improve the environment and the community—for ourselves, our family, our friends, and future generations. By cultivating mindfulness energy, we will gain insight and understanding leading to greater compassion for all beings. It is this compassion that motivates us to take individual and collective action, creating profound change in our communities and societies to improve our well-being and the well-being of our world.

We have a finite amount of energy to spend every day before becoming exhausted. Mindfulness helps you use your energy wisely, spending it on situations, people, and causes that bring you the most joy, meaning, and peace. Mindfulness is the guiding light that already exists inside every one of us. Discover it. We beckon you to use it and let it illuminate your life in every moment. Living like this, you will find yourself savoring your life deeply. Not only will this help you achieve the healthy weight and well-being that you seek; it will bring to the surface life's rich abundance that is so often invisible to us.

PART 1

A Buddhist Perspective on Weight Control

CHAPTER 1

Ending Your Struggle with Weight

> I have been struggling with my weight all my life. I know I have to lose weight. I do not like the way I look. I do not like the way I feel. I have gone on diets, tried diligently to exercise, lost the weight, and had it all come back in no time. I've lost count of how many times I have gone through this yo-yo cycle of loss and gain. I am totally frustrated, ashamed of myself, anxious and overwhelmed about my weight. I am tired of carrying this extra weight around. I do not feel good. Every day is a struggle for me. Every night is a nightmare. I have diabetes now, and I am really worried. I fear that I will not be around to see my children grow up. I am here because I do not want to give up. There must be a way out of this suffering.
>
> **—Participant in a mindfulness retreat**

THIS WOMAN IS NOT alone. Everywhere you turn—from television, magazines, and Web sites to newspapers and radio—you see, read, or hear stories about the U.S. population's frustrating struggle to lose weight. Two out of three adults in the United States are overweight, and one out of three is obese,[1] more than double the rate of obesity in the late 1970s. In scientific terms, we are in the middle of an obesity pandemic, a state of extreme weight gain that is overtaking not only the United States but also much of the globe. This steep rise in obesity over the past thirty years has no parallel in our history, and if we do not change our current trends, the numbers will continue to rise.

And this is largely because our society has become toxic in a way that experts call "obesigenic." We are surrounded by societal forces that drive us to eat more and move less. And the natural result is weight gain, obesity, and the myriad health and emotional problems that go along with them. Yes, it's ultimately a personal decision to eat more than one needs and to not exercise enough, but it's also nearly impossible to escape the pressures around us that lead to unhealthy behaviors. Bombarded day in and day out by unhealthy outside influences, we easily become dissociated from what our bodies truly need and truly want.

Just think of the food court at the local shopping mall. It's a feast of choices that can overwhelm the senses. You see and smell foods that are savory as well as those that are sweet—steak sizzling in teriyaki sauce, oven-fresh pizza, hot cinnamon buns drizzled with snow-white icing, rich coffee confections infused with sugar syrup and topped with cream. The abundance of aromas, colors, and sounds awaken your palate and your urge to eat. In and of itself this isn't necessarily a bad thing—who doesn't love the look and smell of delicious food?—but it often spurs us to eat automatically whether we're actually hungry or not. Before we know it, we've eaten a supersized meal that has two-thirds of the calories we need in one day—and we weren't even hungry to begin with. When this happens day after day, week after week, what began as one enjoyable moment of eating becomes a weight problem that can affect us the rest of our lives. And this is just one of many examples of the impact our surroundings and social networks can have on our weight and health.

In our food supply, there are plenty of foods and drinks that are highly seasoned with salt, sugar, and fat. According to Dr. David Kessler, former commissioner of the U.S. Food and Drug Administration, the food and restaurant industries deliberately produce high-salt, high-sugar, and high-fat foods just so that people cannot resist them and want more. In his book *The End of Overeating: Taking Control of the Insatiable American Appetite*, citing research done in behavioral neuroscience, nutrition, and psychology, he reports that foods high in fat, salt, and sugar alter the brain's chemistry, stimulating the release of dopamine, which in turn is associated with the feeling of pleasure.[2] This is one of the reasons that we crave more of foods and drinks high in fat, sugar, and salt—because they are satisfying.

Advertising is another. It's the market economy's way to shape social norms that propel consumption and profits. And when it comes to the food industry, what they want is for consumers to truly *consume*—to eat as much food and to drink as many drinks as they can stomach, and then have some more. It is telling that in the United States, the food industry spends more on advertising than any other industry except the auto industry.[3] Every day we're exposed to dozens of ads for food and beverages, each one cueing us to eat and drink. And there is no place that is off-limits for eating and drinking: we eat and drink in our cars and at our desks, as we sit in meetings and as we stroll through the shopping malls. It is no wonder that we often find ourselves eating and drinking beyond what we need to satisfy our true physiological hunger. We have created a culture of constant snacking, drinking, and eating.

Now, think of our social norms around physical activity. From the Industrial Revolution in the early 1800s up to our current information technology revolution, we have become increasingly sedentary as we rely more and more on machines, gadgets, and automobiles to do our work and get around. We have drastically diminished the amount of energy we burn every day through bodily movements and the use of our muscles. And with the average home in the United States having more television sets than people,[4] we have become couch potatoes living under the spell of the television.

Together, all these societal forces push us toward eating more calories each day than we expend, and without our being aware of it. Over time these extra calories build up, and before we even realize it we've put on a good deal of weight. And it doesn't take too much for this to happen. Over the course of one year, one hundred extra calories each day—the equivalent of eating one small cookie or of driving a mile instead of walking it—could end up packing ten pounds of extra fat on our bodies.

Given the huge burden these social influences place on us, how can we get back in touch with our bodies and relieve ourselves of the burden and suffering that arise from being overweight? How can each of us reach a healthier weight?

The answer almost certainly doesn't rest with the current weight-loss industry in the United States. Weight-loss programs, diet books, and diet foods, herbs, and pills represent an estimated $59-billion industry in the United States.[5] Thousands of fad-diet books and weight-loss plans come and go. Yet these nearly always fail people over time. You can lose weight on any diet, but there is no scientific evidence that rigid dieting will help you achieve weight loss in the long run. On the contrary, the U.S. population is growing fatter and fatter, and growing increasingly frustrated and discouraged by its failure to lose weight.

Millions are spent on research and development by pharmaceutical companies to find an obesity fix. But there is still no magic pill or formula that can help us lose weight and maintain our lost weight without side effects. The U.S. Food and Drug Administration is cautious in the approval of these fat-fighting drugs. Yet those few drugs on the market that can help people shed a few pounds have unwanted side effects.[6]

The difficult truth is that the basic law of thermodynamics still holds: When we eat more calories than we expend, we gain weight. When we burn more energy through physical activity or exercise than we take in from food and drinks, we lose weight. Though this sounds basic and simple, the fact that so many of us are overweight points to the complexity of the situation. For anyone who has tried many times to lose weight, the thought of trying again may feel like an overwhelming and daunting task. Is it truly possible to change one's habits of eating and moving, especially in the face of a society that pushes us so hard in the wrong direction? How can one begin to make these changes?

The Buddha teaches that change requires insight, and insight cannot begin until we stop and focus our attention on what is happening right in front of us. This stopping, or *shamatha,* allows us to rest the body and the mind. When we have calmed ourselves, we can then go on to look deeply into our current situation. We need to step off our frantic life treadmills, to stop unconsciously doing the same things over and over again that have allowed our weight to creep up. We need to stop, rest, and reflect on a constructive way forward that will end the habits that have led to our current weight issues. We need to be fully aware of what is going on in our daily living. Only then can we begin to change.

Changing Your Habit Energy

There is a Zen story about a man and a horse.[7] The horse is galloping quickly, and it appears that the rider is urgently heading somewhere important. A bystander along the road calls out, "Where are you going?" and the rider replies, "I don't know! Ask the horse!"

This is also our life story. Many of us are riding a horse, but we don't know where we are going, and we can't stop. The horse is our "habit energy," the relentless force of habit that pulls us along, that we are often unaware of and feel powerless to change. We are always running. It has become a habit, the norm of our everyday living. We run all the time, even during our sleep—the time that we are supposed to rest and regenerate our bodies. We are our worst enemies, in conflict with ourselves, and therefore we can easily start conflict with others.

When a strong emotion arises within us like a storm, we are in great turmoil. We have no peace. Many of us try to pacify the storm by watching television or eating comfort foods. But the storm does not calm down after hours of watching. The storm does not go away after a bag of chips or a bowl of ice cream. We hate ourselves afterward for eating the chips and the ice cream. We dread stepping on the scale the next day. We vow to never do it again. But time after time, we do. Why? Because our habit energy pushes us.

How can we stop this state of turmoil? How can we stop our fear, our despair, our anger, and our cravings? We have to learn to become solid and stable like an oak tree, and not be blown from side to side by the emotional storm. We have to learn the art of stopping—stopping our running so that we can be present for and embrace our habit energies of worry, blame, guilt, and fear, and calm the strong emotions that dictate us. We have to learn to live fully in the present moment. We need to practice breathing in and breathing out with all our awareness. We have to learn to become mindful.

When we are mindful, touching deeply the present moment, in the here and now, we gain more understanding, more acceptance, more forgiveness and love of self and others; our aspiration to relieve suffering grows; and we have more chances to touch joy and peace.

We need the energy of mindfulness to recognize and be present with our habit energy so that we may prevent it from dominating us and stop its often destructive course. Mindfulness allows us to acknowledge our habit energy every time it pops up: "Hello, my habit energy. I know you are there." If you just mindfully smile to your habit energy, it will lose much of its strength. The chips stay in the cupboard, the ice cream in the freezer. The storm passes by, and we watch, breathing in and breathing out all the while.

After we become calmer, we can recognize our weight problem more clearly and acknowledge it instead of denying it. This may not be easy for you to do. You may feel angry, frustrated, or fed up about your weight.

Do not suppress these feelings of anger. Instead, as the Buddha has taught us, accept and embrace these difficult feelings, like a mother cradling her crying baby. The crying baby needs the mother's loving care. In a similar manner, your negative emotions and turmoil are crying out loud, trying to get your attention. Your negative emotions also need your tender, loving care. By embracing your negative feelings whenever they arise, you can prevent yourself from being swept away by your emotional storm, and you can calm yourself. When you are calmer, you are more able to see that you already have within yourself the power and the tools to begin to change.

Stopping, calming, and resting are preconditions for healing. If we cannot stop, we will continue on the course of destruction caused by unmindful consumption.

The Four Noble Truths of Healthy Weight

The Buddha offered many teachings to help people end their suffering, the first and most important being the Four Noble Truths. The First Noble Truth is that all of us have suffering in our lives. None of us can escape from it. The Second Noble Truth is that we can identify the causes of our suffering. The Third Noble Truth is that we can put an end to our suffering and that healing is possible. Finally, the Fourth Noble Truth is that there are paths to free us from suffering. We can cultivate our well-being by concretely applying mindfulness to our daily living.

A simple example from the field of medicine can help illustrate the Four Noble Truths. Let's say you are diagnosed with type 2 diabetes (First Noble Truth), which was likely brought on by eating a poor diet and becoming very overweight (Second Noble Truth). Your doctor tells you the situation does not need to be like that and can be controlled (confirming the Third Noble Truth). You follow the doctor's prescription—taking your medicine, eating better, and exercising more—which is your route to healing (Fourth Noble Truth). These teachings of the Buddha originate from a time when suffering was more likely to be caused by a lack of food rather than too much, or by a body overburdened with physical labor rather than one grown ill from lack of use. Yet they apply to all forms of suffering, including those related to being overweight.

Now, let's reflect upon the Four Noble Truths and how they relate to achieving your healthy weight. The self-exploration that begins here and continues throughout this book will help you navigate through all the important factors in your life that affect your weight. It will help you discover what science-based paths you can follow to reach a healthier weight. And through your own awareness, you can discover and decide for yourself what is beneficial and what is not beneficial for your body and well-being.

Through the process, you will realize whether your weight has affected you physically and emotionally. You can become more in touch with the way you have been eating and drinking, the amount of exercise you have been doing or not doing. You can recognize the amount and type of effort you have been spending to control your weight. You can appreciate how your work is affecting your daily lifestyle and your weight. Through all these reflections, you can gain insights from your past that can lead you to success on your path of healing.

As you read the rest of this chapter and this book, read with an open mind and an open heart. Do not struggle with the concepts—the information here is not intended to simply add to your warehouse of knowledge. Be like the earth. When the rain comes, the earth simply opens up to the rain and soaks it all in. Allow the wisdom in this book to nurture the seeds that lie deep in the soil of your consciousness so that they can sprout and mature into the transformative energies of mindfulness and insight. A teacher cannot give you the truth. The truth is already in you. A teacher can only offer you the chance to awaken your true self.

Enlightenment, peace, and joy will not be granted by someone else.
The well is within us,
And if we dig deeply in the present moment,
The water will spring forth.[8]

The First Noble Truth: Being Overweight or Obese Is Suffering

When we are overweight, every part of the body may feel the burden. Our knees may ache, carrying too much weight, becoming swollen and stiff with arthritis.[9] Our heart may labor harder, our blood pressure may rise, and harmful plaque may build up inside the lining of the arteries, heightening the risk of heart attack and stroke. Our breathing itself may become a problem around the clock as the risk of asthma, chronic obstructive pulmonary disease, and sleep apnea increases.[10]

Many of these increased health risks exist in people who are merely overweight, not just among those who are obese. Diabetes, an insidious disease with disabling and deadly complications, is two to four times more likely to strike someone who is overweight than someone who is at a healthy weight, and it is five to twelve times more likely to strike someone who is obese.[11] The risk of cancer in several parts of the body—breast, colon, esophagus, kidney, pancreas, and uterus—is higher in people who carry excess body fat than in people who do not.[12] So is the risk of having gallstones that require the gallbladder to be removed.[13] The risk of infertility,[14] cataracts,[15] and perhaps even dementia may be higher in people who are obese than in those who are at a healthy weight.[16] Given the toll that excess weight takes on all areas of the body, it is no surprise that being overweight or obese in midlife increases the chances of dying early.[17] Even people who are not overweight but have gained more than twenty pounds since the start of college have an increased risk of dying early.[18]

Those who are overweight and obese suffer in countless other ways as well, due to the pervasive stigma associated with weight.[19] As children they may, owing to their weight, become the target of teasing and bullying from their peers. As adults, they may be less likely to win a job or a promotion, or they may be stereotyped as lazy or less disciplined. Even in the doctor's office, they may face prejudice because of their weight.

What type of suffering have you endured because of your weight? Physical pain? Emotional pain? A feeling of shame, insecurity, regret, anger? Identifying and acknowledging the nature and depth of your suffering may be difficult. You may want to suppress it and not deal with it. However, our first step toward healing and transformation is to recognize the existence of our suffering and not run away from it.

The Second Noble Truth: You Can Identify the Roots of Your Weight Problem

Before you can change your weight, you must have a better understanding of the reasons that you are overweight. Fundamentally, weight gain or loss results from changing the balance of energy coming in (the calories we eat and drink) and going out (the calories we burn off through our everyday activities). Yet science has found that many factors can lead us to get out of balance and gain weight—our ancestry, our lifestyle choices, and our surroundings.

Take the time to reflect on the numerous factors that we describe here. Look deeply to see whether they apply to you so that you may understand the true nature of your problem with weight. Looking deeply requires courage. The causes are knowable, and with diligent effort you can get to the bottom of them. With greater insight into the reasons that you are overweight, you can begin to determine what course of action you can take to achieve a healthier weight.

Know that the attachment to pleasurable desires can cause us to suffer. As we crave insatiably for delightful and pleasurable experience through our consumption of foods and drinks and our sedentary way of life, we are well on our way to gaining weight. Do these desires really satisfy you in the long run and bring you happiness? Not likely, since all these are temporary fixes that get us to gain more weight. When you succumb to these cravings, you are perpetuating the cycle of frustration, anxiety, and suffering.

Buddhism describes creatures known as *pretas,* or Hungry Ghosts, who have insatiable appetites for food, drinks, and other cravings. They are desperate beings who are always hungry, with tiny mouths; long, narrow necks; and distended bellies. Though they are constantly ravenous, driven by the desire to eat, their tiny mouths and necks prevent them from swallowing the food they ingest. The act of eating does not help them overcome negative emotions and cravings. Eating more only causes them to have more pain and agony. Are you consuming like a Hungry Ghost?

As you begin to look deeply into the roots of your weight problem, take care not to be harsh on yourself. The "judge" inside your head often makes you feel bad about all the "shoulds"—you should not have eaten that cheesecake, you should have spent more time at the gym. You may also be daunted by your past failures and struggles with weight. It is time to stop blaming yourself for these failures. Perhaps you were following the wrong advice. Perhaps you were able to lose some weight initially on one diet or another, but the diets were too restrictive, your cravings took hold, and you eventually gave up and gained the weight back. You are not separate from your family and environment. In the past you did not have enough of the right conditions supporting you to maintain a healthy weight.

Do you understand why you did not succeed? What were your obstacles? Do not get lost in regret about your past mistakes. The past is the past. It is not the present. You can seize the present moment—any present moment—to begin anew. Just as you embrace your negative feelings, embrace your weight problem like a mother cradling her crying baby, so that you can transform your fear, despair, anger, frustration, and self-criticism. Mindfulness practice can help you become calmer, so that you can look at your situation in a more detached way, without self-condemnation. This frees you to focus on the solutions rather than dwell on the past or your problems. The Buddha said that if we know how to look deeply into our suffering and recognize what feeds it, we are already on the path of emancipation.

Do Your Parents Have Weight Problems?

What you have inherited from your ancestors' gene pool can affect your weight. Studies have shown that when one of a child's parents is overweight, the child is more than twice as likely to become an overweight adult, regardless of whether the child herself is at a normal weight.[20] Having two overweight parents further increases the chance that the child will become overweight.[21] Nevertheless, parental influences on our weight could be due to nature, nurture, or a combination of both. When we were young, parents also controlled what and how much we ate as well as how active we were. If your mother breast-fed you, it may have lowered your risk of becoming overweight.[22] If your father encouraged you to "clean up your plate" as a habit, you may find it really hard to stop eating today's supersize portions even though you are full. Did your parents spend their free time playing with you in the yard, or did you spend much of your family time slouched together in front of the television set? As you consider the role of parental influences on your weight, keep in mind that your genetics are not your destiny—and genetics alone cannot explain the rapid rise in obesity we have seen over the past thirty years. Even if your parents were overweight, you can still achieve a healthy weight by following a healthy lifestyle. It just means that you may have to pay more attention to what you eat and how much you move than someone who does not have a genetic tendency to be overweight.

Do You Drink Too Much Sugary Soda?

Drinking sugar-sweetened beverages can contribute to weight gain. A study in teens found that for every additional can of soda they drank a day, their chances of becoming obese increased by 60 percent.[23] The Nurses' Health Study found that women who increased their consumption of sugary drinks from one or fewer drinks per week to one or more drinks per day gained more weight over a four-year period than women who cut back on sugary drinks.[24]

Scientists believe there are several reasons that drinking sugary drinks contributes to weight gain. The calories from soda are often "invisible." When you drink your calories rather than eat them, you may not cut back on other foods to compensate for the liquid calories. With a twelve-ounce can of sugary soda containing nearly ten teaspoons, or about 150 calories, of sugar, it's easy to stack up extra calories during the day, especially if you're drinking sugary soda to quench your thirst. Soft drinks may also increase your sense of hunger or decrease your sense of satiety or fullness.[25]

As you consider the role of sweetened drinks on your weight, notice whether you feel hungrier after you drink a sugary soda. If you have already cut back on sugary drinks and substituted diet ones, notice whether drinking the intensely sweet sugar substitutes has conditioned your palate to expect, crave, and seek supersweet foods.

Do You Get Less Than a Half Hour of Exercise or Physical Activity Each Day?

For weight control, the "energy out" side of the energy-balance equation is just as important as the "energy in." There's strong evidence that getting enough physical activity can help prevent weight gain and, when combined with a lower-calorie eating plan, can help promote weight loss. What is enough? It depends on how fit you are. For some people, taking a brisk half-hour walk five days a week would be enough. For others, taking a high-intensity spinning class for seventy-five minutes a week would be enough. For people who are very inactive, just getting moving is a start. We discuss physical activity for weight control and for general good health in chapter 6, "Mindful Moving." For now, reflect on whether you get enough physical activity. If not, why not?

Do You Watch More Than One Hour of Television a Day?

Many studies in adults and children have shown that watching too much television increases one's risk of becoming overweight. The Nurses'

Health Study, for example, found that for every additional two hours of television that women watched per day, their risk of becoming obese increased by 23 percent. Even getting a lot of physical activity did not fully protect these women from the effects of TV watching on weight: among the most active women, those who watched more than twenty hours of television per week had a higher risk of becoming obese than those who watched less than six hours of television per week.[26]

Researchers believe there are several possible ways that watching too much television could lead to weight gain. Sitting around and watching television may take the place of more physically demanding activities, so the "energy out" side of the energy balance equation goes down. TV watching may also affect the "energy in" side of the equation: people tend to eat *while* they watch television and also tend to eat *what* they watch on television—fast food, sugary drinks, and other high-calorie snacks. This adds up to extra calories in, fewer calories out, and, ultimately, to weight gain. As you reflect on your TV-watching habits, think about why you spend as much time as you do watching television. Do you watch a lot of television to avoid boredom? To avoid communicating with your family members? Or to cope with stress? What other activities might you do instead? (See a list of suggestions in appendix D for ideas.)

Do You Get Enough Sleep?

A good night's sleep is essential for good health. New research suggests that a good night's sleep may also be essential to controlling your weight.[27] The Nurses' Health Study, for example, followed seventy thousand women for sixteen years. Women who skimped on sleep—getting five hours or less each night—were 15 percent more likely to become obese than women who got seven hours of sleep per night.[28] Scientists are still teasing out why lack of sleep may lead people to pack on the pounds. People who don't get enough sleep may be more fatigued than people who get a healthy night's sleep,[29] decreasing the "energy out" side of the energy-balance equation. Or staying awake for a longer period of time may simply give people more opportunities to eat, increasing the "energy in" side of the energy-balance equation.[30] Sleep deprivation may shift the balance of key hormones that control appetite, making sleep-deprived people hungrier than people who get enough sleep.[31] One small study found that sleep-deprived volunteers reported more hunger, especially for high-carbohydrate and high-calorie comfort foods.[32] If you are not getting enough sleep, think about why. Are you lying awake at night filled with anxiety? Do you stay up too late watching television? Do you notice that you are hungrier on days when you have not gotten as good a night's sleep?

Do You Eat Mindlessly?

Nowadays, with all the societal pressures and the "high speed" living of our Internet age, much of our eating happens on autopilot. We do not pay attention to how much food is served or how much we have eaten, how tasty the food is or whether we're even hungry at all. Instead, how much we eat is often driven by external cues—the size of the bowl, the size of the plate, the portion size of the food itself. Given the supersizing trends over the past twenty years, it is easy to fall victim to "portion distortion" and to lose sight of how much food is an appropriate amount to eat.[33] Recently, researchers have conducted scientific studies looking at how mindless eating affects our food consumption. What they found is that mindless eating can easily lead to overeating. In a classic experiment, people at a movie theater were served fresh or stale popcorn in different-size containers. Moviegoers who were given stale popcorn said the taste was "unfavorable." Yet when they were served stale popcorn in a large container, they ate 61 percent more popcorn than they did when it was served in a small container while they were watching the movie—and they underestimated the amount of popcorn they ate.[34] In another experiment, graduate students at a Super Bowl party who served themselves from large bowls ate 56 percent more snack food than students who served themselves from smaller bowls.[35]

The larger the portion size, the less able we are to estimate how many calories we are eating.[36]

The cues for mindless eating reach beyond the size of a plate or the size of a portion. Our whole surroundings support mindless eating, from the ads on TV to fast-food "dollar menus" to favorable placement of unhealthy foods on supermarket shelves. All these cues combined can make it very, very difficult to find what it is our bodies truly need. Are you often eating on the run, in the car, or at your desk? Do you have to dine out often because you have not had time to cook? And do you find yourself making unwise food choices when eating out?

Practicing mindfulness can help us avoid the external cues that trap us, avoid mindless eating, and focus in on the practices that keep us healthy.

Mindlessness is the opposite of mindfulness. Eating is not the only activity that we do mindlessly, and we are driven to mindlessness by more than just the size of a bowl or plate. We drink a cup of tea and focus more on the worries and anxieties of the day than on living the moment of enjoying the tea. We sit with someone we love, and rather than focus on the person and this moment we have with them, we're distracted by other thoughts. We walk but are more focused on reviewing the talking points for our next appointment than on the serene moment we're having as we walk. We are usually someplace else, thinking about the past or the future rather than the now. The horse of our habit energy is carrying us along, and we are its captive. We need to stop our horse and reclaim our freedom. We need to shine the light of mindfulness on everything we eat and do, so the darkness of forgetfulness will disappear.

Do You Live or Work in an Environment That Makes It Difficult to Eat Healthfully and Keep Moving?

Where you live and where you work can have important implications for whether you can eat well and stay active. If healthy choices are not available in your workplace or neighborhood, it makes it very difficult for you to eat well, no matter how knowledgeable or determined you are. If your neighborhood is not safe enough for you to walk, jog, or ride a bike in, it will deter you from staying active. Pay attention to your surrounding environment, and take note of the various barriers to active living and healthy eating. What prevents you from following your good intentions? Are your sincere efforts being sabotaged by family or friends? Does your job or what you do on a daily basis prevent you from staying on course with healthy eating and active living? Do you have too much work-related stress? Do you have time for yourself?

As you start to clear your mind of distractions, these barriers will become clearer to you, and you will start to work on ways around them as well as ways to work with people in your community to make a healthier environment for everyone.

Other Factors Feeding Your Weight Problem

Scientific and public health experts are working hard to figure out what can be done to turn the obesity epidemic around. Yet, science still does not have all the answers about what leads people to gain too much weight. So it is important to consider other factors that may have caused your weight problem. It may be helpful for you to reflect on the following questions about your own attitudes, thoughts, feelings, and actions that may have led you to eat more and move less. Be honest with yourself. Write your reflections in a journal so that you can review them later and gain a better understanding of yourself. Once you are conscious of these attitudes, thoughts, feelings, and actions, you can work, step-by-step, to change them—to break the mindless forces of habit that have led you to eat more and move less.

How do you feel about your current weight? Is having a healthy weight a high priority in your life that's worth your time and energy to address? Do you have enough concentration to focus on your weight problem, your poor eating habits, and your sedentary lifestyle? What is distracting you from your focus?

Do you feel that you are doomed to be overweight and it does not matter how hard you try? Do you eat just to feel better, if only for short time?

Are you eating before going to bed because you are tired? Are you attached to certain unhealthy foods late at night? Why? What triggers you to eat again just a little while after finishing your dinner?

Are you using food to fill an emotional void, relieve loneliness, or cope with your anxiety, fears, or stress? Do you continue to eat when you are full, and how does this make you feel? Is overeating a Band-Aid to cover up another type of pain? Are you trying to feed an emotional hunger? Are you using food as a crutch? Do you use food to ward off the painful, heart-wrenching feelings that you have buried deep in your heart? Reflect mindfully about how you may use food to cope with your negative emotions; you will see that edible food is not the right nourishment for them.

Where do you get your information? From a credible source? Or from magazines, TV shows, or advertisements that sensationalize results and make promises they can't fulfill? Do you find yourself being affected by food commercials on television or in magazine advertising? Have you been the victim of countless diet books that have led you to lose confidence in your ability to reach a healthy weight?

Do you have a preexisting health condition that prevents you from exercising? If you do, did you seek out the advice of a professional to help you find appropriate physical activities that you can do on a daily basis?

Listen to your heart. What are your inner longings? How are you going to fulfill these longings? Do your self-talk, beliefs, or interactions with others hinder you from maintaining healthy eating and active living? Are you your own worst enemy? Have you put time and effort into healthy eating and active living? If not, why not?

THESE ARE ALL VERY complex questions that take time, effort, and sometimes a bit of painful soul-searching to answer, but they are very important issues to address. And tackling them head-on and resolving them is just as important to reclaiming our healthy weight as the better-studied issues, like getting more exercise and cutting back on sugary soda.

As you learn to be more mindful, to better focus on what is happening in the present moment, the barriers and motivations that drive you toward unhealthy habits will become clearer, as will the path away from them toward better health.

The Third Noble Truth: Reaching a Healthy Weight Is Possible

You can put an end to your weight problem. You have already taken the first step: by spending energy to understand the roots of the problem, you have stopped running away from it. When you direct your attention to your excess weight and the suffering associated with it, you can see the potential for wellness. You also understand that it is possible to reach your goal of a healthy weight when you take the right actions. Remember, there was a time before you became overweight when your weight was normal. It is easy to forget that.

Ask yourself where you are in this healthy-weight journey. Focus on whether or not you aspire to be a different person, to feel better about yourself, to be able to function better, to be happier. Ask yourself what this extra weight means to you, and ask yourself whether you are truly ready to let it go.

To be successful, it is very important for you to believe that you can achieve a healthy weight. Believing in yourself, having the faith that you can change the habits that do not serve you well, and adopting science-based wisdom are important for successful transformation of our behavior. According to psychologist Albert Bandura, "perceived self-efficacy" is essential for any behavior change.[37] Self-efficacy is simply the belief that one *can* carry out a behavior necessary to produce a desired outcome. What we believe can significantly affect what we can achieve. People who believe that they can reach a healthier weight through healthy eating and active living set relevant goals that they perceive to be important for the desired change. They believe that those goals are attainable, and they believe that they have the ability to carry them out.

What are your current beliefs? Are they real, or are they shadowed with illusions from your past experiences, failures, and disappointments? The past is the past. The past is your teacher and can offer valuable lessons on what worked and what did not work for you. But it is not your present reality. It remains your present reality only if you allow it to be. Do not let your past experiences hold you back. Your failures do not need to determine your current or future experience. Focus on the present. When you focus on the present, you do not give any power to your past actions.

Recognize that losing weight *is* possible and that you *can* do it. It will not be as easy as popping a pill and watching the pounds melt away, but it will be a journey—sometimes hard—that will be well worth taking.

Learn to listen to your body with compassion. You are more capable of helping yourself if you treat yourself with compassion and a loving heart. Your guilt and hopelessness will diminish, and you will be more accepting of yourself. Don't judge yourself harshly. Love yourself, and affirm that you can reach a healthy weight. You are losing the weight for yourself, not to please anyone else.

As you start to recognize your innate capacity to be more mindful, you will become calmer, and it will be easier to find solutions to the problems you face. Allow yourself to truly feel what your life would be like if you could maintain a healthier weight. Affirm to yourself that reaching a healthier weight is possible when you pay attention to and take care of your weight. When you take care of an orchid plant, the plant needs your focused attention, to be watered and nurtured on a regular basis. Without such care, the flowers will wither and die prematurely and you cannot enjoy their beauty. You are just like the orchid plant. You need tender loving care for your ideal to be realized. When we look at all beings, including ourselves, with eyes of love and compassion, we can take care of ourselves better. With mindfulness, we can nurture ourselves with greater ease and interest, and our effort will come more naturally.

Even if you have been burdened and tormented by your weight throughout the years, there are seeds of well-being inside you. But you may have lost sight of this because the discomfort of your overweight is overpowering. When you have a toothache, you call your dentist and ask for an emergency appointment to relieve your pain. You know deeply at that point that not having a toothache is happiness. Yet later, when you don't have a toothache, you forget and do not treasure your non-toothache. Practicing mindfulness helps us appreciate the well-being that is already there and realize that further well-being is possible if we take the right actions.

We need to water the seeds of joy in ourselves in order to realize well-being, including the well-being that comes from being at a healthier weight. Please ask yourself, "What nourishes joy in me? What nourishes joy in others? Do I nourish the joy in myself and the joy in others enough? Do I appreciate the many reasons for joy that are already in my life? Or have I been living in forgetfulness, taking many things for granted?" If you have good eyesight, appreciate it, even though it's easy to forget what a gift this is. If you're able to sleep well, appreciate this. If your lungs are healthy, appreciate the simple fact that you can breathe in and out easily. The same goes for myriad things we can do every day but don't take note of.

When you suffer, you can look deeply at your situation and find the many blessings that are already surrounding you. It is wonderful to sit with a pen and paper and write down all the conditions for happiness that are already there, already available to you right in this moment.

When you do this, you have a firm foundation from which to better embrace and transform your suffering. Transforming our suffering is like becoming an organic gardener, one who does not discard the unwanted scraps from the kitchen or the yard. Instead, the gardener composts these scraps so that they can nourish the flower. You can transform the unwanted garbage in you—your depression, fear, despair, or anger—into the nourishing energy of peace and joy. Don't throw away or deny your suffering. Touch your suffering. Face it directly, and transformation is within your reach.

Negative habits can be changed. You *can* begin anew. Try to be fully aware of your inner motivations for wanting to reach your healthy weight. Why do you want to lose weight? Allow yourself to truly feel that life would be significantly better without your weight problem, as you would feel better and become healthier. Your intention to reach a healthy weight has to come from you—and from nobody else.

You need to get back to the wisdom of balance and moderation. Think of balance and moderation as seeds that lie dormant in your consciousness. Water them so that they can grow and be strong. Attaining a healthy weight is your choice. And it is a practice, not an idea.

The Fourth Noble Truth: You Can Follow a Mindful Path to a Healthy Weight

The mindful path to a healthy weight is not a diet that you go on and off. It does not rely on any pills or potions. All it takes is your belief and affirmation that you can follow a mindful path, and your willingness to commit to this path. Following a mindful path means creating your own personal goals for healthy eating and physical activity, goals that you believe you can carry out and live with, day after day. These goals must be realistic for your own life's demands. Over time, as you progress on this mindful path, it becomes your way of living, one that allows you to reach your healthier weight with great ease and confidence.

Appreciate the lessons that your excess weight is offering you. Your excess weight is like a bell whose sound reminds you that your past actions, your past way of living, did not serve you well. You can liberate yourself from the imprisonment, struggle, and burden of your weight. Though you are not solely responsible for your current state, you are the only one who is able to change it. You must act for yourself. No one else can do this for you. The first step is to be aware that *you* are choosing to change.

Bear in mind that everything is impermanent, including your extra weight. The mindful path to healthy weight is to start living with mindfulness, which will help you be more conscious of what you are thinking, seeing, hearing, feeling, eating, and doing throughout the day.

Take small steps toward change. Do not aim for unrealistic goals that require a quantum leap. This usually sets you up for a vicious cycle of failure. Another experience of failure is not what you need. Chances are that your past failures in weight-loss attempts have reinforced the notion in you that you cannot do it.

By accomplishing little steps, you start to have a taste of what you can do. You begin to have a taste of success. Every day, allow yourself to witness that it is possible for you to make changes in the right direction, no matter how small. Success breeds success. It is contagious.

Set a realistic goal for weight loss. For most of us, this means losing one to two pounds a week. Losing weight steadily allows your new, healthy habits to be firmly rooted. When you lose weight quickly with fad diets, you usually regain the lost weight in a short time. The best approach is to establish healthy eating and physical activity habits that you can stick with.

There is a National Weight Control Registry of more than five thousand women and men who have each lost more than thirty pounds and kept it off for at least a year.[38] Most of them lost the weight on their own. How did they do it? They exercised and burned an average of four hundred calories a day, which is about sixty minutes of brisk walking. They ate less, keeping to about fourteen hundred calories a day. Plus, they watched less television and limited their visits to fast-food restaurants. There are other examples of success, from a fast-food chain's formerly overweight pitchman to people who participated in reality shows like *The Biggest Loser.* The key is to be able to stick with eating well, eating less, and moving more.

Learning what to do to lose weight is not difficult. Putting that knowledge effectively into practice, though, is the key challenge. Despite advances in weight management research over the past decades, people worldwide still struggle to make the lasting lifestyle changes that would help them achieve and maintain a healthier weight.

So, what can help you begin to make healthy behavior changes and keep them up over time?

Mindfulness.

Mindfulness is a way of living that has been practiced over twenty-six hundred years by millions of people to help them transform their suffering into peace and joy. Applying mindfulness to your suffering with weight gives you a catalyst that you can draw on at will to change your behavior. Consider mindfulness as your ally to help you get out of your own way, change your habits that are counterproductive, and overcome the obstacles and difficulties that led you to be overweight.

Start first with practicing mindful breathing, mindful eating, and mindful walking every day. We will explain these practices in chapters 4, 5, and 6. It may seem like a lot to start with, but it doesn't take much at this stage. We are already breathing as long as we are alive, and we are eating and walking every day. Mindfulness is simply a different way of breathing, eating, and walking.

Don't be discouraged if you find yourself not being able to follow your plan 100 percent. As long as you are heading in the right direction, you are making progress. Be patient with yourself. Even if you change only one unhealthy habit a week, you will be making twenty-six changes in six months and fifty-two changes in one year. Act with determination. Practice diligently. Take one moment at a time.

In addition to your personal mindfulness, you will need to look outside yourself if you're to reach a healthy weight. Besides your own will and action, you also need support at home, at work, and throughout your community to be able to eat well and stay active. If healthy choices are not available at work, it will be much more difficult for you to eat a healthy lunch. If your community does not have any food outlets besides convenience stores that do not stock many healthy choices, it will be hard for you and your family to eat well.

It is extremely challenging to change on your own. Build a support community or *sangha* (the Sanskrit word for a Buddhist community of spiritual practitioners) around you to help you stay on track. The support system can be made up of people you meet with in person or online. Think of all the people you come into contact with in your life. Which can be your potential allies, offering you support in the myriad of small daily actions? How about your family members, your friends, your coworkers, and your health-care team? How about using the Web and other tools to remind you to eat, move, and live mindfully? See appendix A to download the mindfulness bell sound to your computer to remind you to come back to the present moment. You may also want to have a digital watch or cell phone that you can program to beep at regular intervals, say, every hour. The beep can be a reminder for you to stop whatever you are doing and breathe deeply three times. In all the practice centers in the tradition of Plum Village (Thich Nhat Hanh's monastery in France), whenever the phone rings or the clock chimes in the dining hall, people stop everything they are doing and breathe consciously, releasing all thinking and any tension.

Ending Your Struggle with Weight: The Path Begins Here

You have within you the wisdom, the strength, and the ability to follow through with your plan to lose weight. Through reading this book, you will learn scientific facts that will help you eat more healthfully and be more active. You will learn more about the internal and external obstacles that may have prevented you from reaching your healthy weight in the past. You will know yourself much better than before. You will see whether your thinking is working for you or against you. You will realize that what you do every day for leisure and for work affects your weight and well-being. You will be more conscious of how your level of focus, the extent of your mindfulness or mindlessness, and your commitment are all affecting how much you weigh.

Make a serious pledge to yourself. Begin writing a personal healthy-weight mission statement. This mission statement is a symbol and reminder of your commitment and can help you see more clearly what you intend to achieve. As you go through this healthy-weight journey, return to your healthy-weight mission statement periodically to help you focus, get inspired, renew your ongoing commitment, and stay on track. You may also like to post it where you can read it often.

Your mission statement for healthy weight and well-being should be made up of concrete, achievable goals. It should begin broadly with what you want to achieve overall and then have some more specific goals you can work toward along the way to your ultimate goal. For example:

My Mission Statement for Healthy Weight and Well-Being

Through being more mindful and paying better attention to my health and well-being, I will lose 25 pounds by _________, *(insert date)* and keep it off over the following year and beyond.

Beginning Goals, _________ *(insert date)*

- I will practice some level of mindfulness every day, with a goal of increasing the amount every week.
- I will walk at least 5,000 steps (measured by a pedometer) or half an hour every day, and slowly increase this each week until I reach 10,000 steps or one hour per day.
- I will buy more fruits and vegetables.
- I will not buy any sugary sodas.

Revised Goals, _________ *(insert date)*

- I will work on being mindful for at least two hours a day, with a goal of increasing the amount every week.
- I will walk at least 10,000 steps every day.

- I will buy more fruits and vegetables.
- I will avoid all fast food.
- I will not buy any sugary sodas.

Mission statements are works in progress, just like ourselves. While the overall goal of your mission statement will likely stay constant over time, the smaller goals will change and be revised as you gain experience and figure out what you've been successful with and what you still need to work on.

As you think about your mission statement and the commitment you've made to yourself, you can sometimes become overwhelmed by the larger goals. But remember that those are the goals to be reached much further into your journey. Right now, just stay focused on the individual steps that will get you moving toward your goals. Instead of focusing just on the scale and your weight goal, stay focused on your daily mind-set and actions moment to moment. Smile to yourself, and feel good about every change that you can make, no matter how small. It takes time and determination to transform the deeply entrenched habits that caused you to gain weight.

You are not expected to be able to change overnight. Just like a woodpecker that has to keep pecking away to create a hole in the tree trunk, you need to focus and keep practicing. Stay present in the now so that you are fully aware and can take concrete steps in the direction of healthy eating and active living. To stay on this route to health and well-being, you need to wake up from your autopilot mode. You have to live deeply and with more awareness so that you can be attentive to each moment. Mindfulness practice is the key that will help you free yourself from unconsciousness and forgetfulness. Mindfulness can help you eat, move, and live more consciously.

In the beginning, it maybe difficult to change your daily routines. As you bring more mindfulness into your daily living gradually and consistently, you will increase your awareness of your daily activities. Before you fully realize it, mindful living will become your new habit, part of your daily being.

A bonus of practicing mindful living, as many others practicing mindfulness throughout the centuries have found, is the sense of being more centered, more joyful, more at peace with yourself. You will wonder why you have lived so much of your life in numbness and forgetfulness. This numbness prevented you from being fully conscious of what caused your suffering and undesirable states of mind and body. This forgetfulness caused you to miss your appointment with life, out of touch with all its beauty and wonders.

By combining science-based advice with the practice of mindfulness, we have the tools to transform the unhealthy habits that have led to our current weight. Our mindfulness will lead us to adopt lifestyle choices that are not only good for ourselves but also good for our planet. We will realize that we cannot attain well-being all by ourselves. Our well-being is intimately dependent on the well-being of others. Our health is dependent on the health of our planet, and the health of our planet depends on us. We all need to consume and act in such a way that the health of our planet is preserved for our children, our grandchildren, and all future generations. We cannot focus just on ourselves if a future is to be possible. We need to maintain the well-being of all. Each of us can contribute by mindfully taking care of ourselves—and our home, planet Earth—for our generation and for generations to come.

CHAPTER 2

Are You Really Appreciating the Apple?

An Apple Meditation

The apple in your hand is the body of the Cosmos.

—Thich Nhat Hanh

LET'S HAVE A TASTE of mindfulness. Take an apple out of your refrigerator. Any apple will do. Wash it. Dry it. Before taking a bite, pause for a moment. Look at the apple in your palm and ask yourself: When I eat an apple, am I really enjoying eating it? Or am I so preoccupied with other thoughts that I miss the delights that the apple offers me?

If you are like most of us, you answer yes to the second question much more often than the first. For most of our lives, we have eaten apple after apple without giving it a second thought. Yet in this mindless way of eating, we have denied ourselves the many delights present in the simple act of eating an apple. Why do that, especially when it is so easy to truly enjoy the apple?

The first thing is to give your undivided attention to eating the apple. When you eat the apple, just concentrate on eating the apple. Don't think of anything else. And most important, be still. Don't eat the apple while you are driving. Don't eat it while you are walking. Don't eat it while you are reading. Just be still. Being focused and slowing down will allow you to truly savor all the qualities the apple offers: its sweetness, aroma, freshness, juiciness, and crispness.

Next, pick up the apple from the palm of your hand and take a moment to look at it again. Breathe in and out a few times consciously to help yourself focus and become more in touch with how you feel about the apple. Most of the time, we barely look at the apple we are eating. We grab it, take a bite, chew it quickly, and then swallow. This time, take note: What kind of apple is it? What color is it? How does it feel in your hand? What does it smell like? Going through these thoughts, you will begin to realize that the apple is not simply a quick snack to quiet a grumbling stomach. It is something more complex, something part of a greater whole.

Then, give the apple a smile and, slowly, take a bite, and chew it. Be aware of your in-breath and out-breath a few times to help yourself concentrate solely on eating the apple: what it feels like in your mouth; what it tastes like; what it's like to chew and swallow it. There is nothing else filling your mind as you chew—no projects, no deadlines, no worries, no "to do" list, no fears, no sorrow, no anger, no past, and no future. There is just the apple.

When you chew, know what you are chewing. Chew slowly and completely, twenty to thirty times for each bite. Chew consciously, savoring the taste of the apple and its nourishment, immersing yourself in the experience 100 percent. This way, you really appreciate the apple as it is. And as you become fully aware of eating the apple, you also become fully aware of the present moment. You become fully engaged in the here and now. Living in the moment, you can really receive what the apple offers you, and you become more alive.

By eating the apple this way, truly savoring it, you have a taste of *mindfulness,* the state of awareness that comes from being fully immersed in the present moment. Letting go for those few short minutes and living in the here and now, you can begin to sense the pleasure and freedom from anxiety that a life lived in mindfulness can offer.

In today's world, mindless eating and mindless living are all too common. We are propelled by the fast pace of high-tech living—high-speed Internet, e-mails, instant messages, and cell phones—and the expectation that we are always on call, always ready to respond instantly to any message we get. Thirty years ago, hardly anyone would have expected to receive a reply to a phone call or letter within the same day. Yet today, the pace of our lives is utterly harried and spinning out of control. We constantly have to respond to external stimuli and demands. We have less and less time to stop, stay focused, and reflect on whatever is in front of us. We have much less time to be in touch with our inner selves—our thoughts, feelings, consciousness, and how and why we have become the way we are, for better or worse. And our lives suffer because of it.

Some of us find that it is too inconvenient and difficult to eat a whole apple. So major food outlets now sell "value-added" apples—presliced apples, packed in bags and coated with an all-natural flavorless sealant so that they won't turn brown or lose their crispness for up to three weeks. These apples epitomize the new food-marketing concept of "snackability":[1] There are no crumbs and no fuss, nothing to interrupt the repetitive movement of hand to bag and food to mouth. Aside from the inherent lack of freshness in these "snackable" precut apples, they also promote mindless eating—in the car, in front of the TV, at the computer, whenever and wherever. And while there are certainly much less healthy snack foods than precut apple slices, the pattern of eating is one we all experience and that food marketers promote with a vengeance.

Most of the time, we are eating on autopilot, eating on the run, eating our worries or anxieties from the day's demands, anticipations, irritations, and "to do" lists. If we are not conscious of the food we eat, if we are not actively thinking about that apple, how can we taste it and get the pleasure of eating it?

Eating an apple mindfully is not only a pleasant experience; it is good for our health as well. The adage "An apple a day keeps the doctor away" is actually backed by solid science. Research shows that eating apples can help prevent heart disease because the fiber and antioxidants they contain can prevent cholesterol buildup in the blood vessels of the heart. The fiber in apples can also help move waste through the intestines, which can help lower the risk of problems such as irritable bowel syndrome. Eating the apple with the skin—especially when it is organic—is better than eating it without the skin, as half of the vitamin C is under the apple's skin; the skin itself is rich in phytochemicals, special plant compounds that may fight chronic disease. Apples are also packed with potassium, which can help keep blood pressure under control.

Beyond the health benefits and pleasure an apple can provide, when we view the apple on an even grander scale we can see it as a representative of our cosmos. Look deeply at the apple in your hand and you see the farmer who tended the apple tree; the blossom that became the fruit; the fertile earth, the organic material from decayed remains of prehistoric marine animals and algae, and the hydrocarbons themselves; the sunshine, the clouds, and the rain. Without the combination of these far-reaching elements and without the help of many people, the apple would simply not exist.

At its most essential, the apple you hold is a manifestation of the wonderful presence of life. It is interconnected with all that is. It contains the whole universe; it is an ambassador of the cosmos coming to nourish our existence. It feeds our body, and if we eat it mindfully, it also feeds our soul and recharges our spirit.

Eating an apple consciously is to have a new awareness of the apple, of our world, and of our own life. It celebrates nature, honoring what Mother Earth and the cosmos have offered us. Eating an apple with mindfulness is a meditation and can be deeply spiritual. With this awareness and insight, you begin to have a greater feeling of gratitude for and appreciation of the food you eat, and your connection to nature and all others in our world. As the apple becomes more real and vibrant, your life becomes more real and vibrant. Savoring the apple is mindfulness at work.

And it is mindfulness that will help you reconnect with yourself and become healthier in mind, body, and spirit now and in the future.

CHAPTER

You Are *More* Than What You Eat

> When something has come to be, we have to acknowledge its presence and look deeply into its nature. When we look deeply, we will discover the kinds of nutriments that have helped it come to be and that continue to feed it.
>
> **—Samyutta Nikaya 2, 47**

ADVANCES IN SCIENTIFIC RESEARCH since the late twentieth century reinforce the understanding that our bodies affect our minds and our minds affect our bodies. Skipping breakfast has been shown to dull students' memories and lower their test scores.[1] And physical activity has been shown to sharpen thinking, reduce anxiety, reduce depression, and enhance memory.[2] The mind also has a powerful effect on our bodies. In stressful situations—like being approached by a bear in a national park—our brains kick in the fight-or-flight response to help us cope with the threat. Our sympathetic nervous system is recruited, stimulating the release of glucagon and cortisol hormones to increase fuel to our muscles and our brain, helping us to make the right decision and escape danger. More recently, research has found that the risk of heart disease increases in both men and women as levels of anxiety, anger, or more general symptoms of distress increase.[3] Therefore, to attain well-being, we need to take care not only of our bodies but also of our minds. Mindfulness practice is central to seeing the interdependence of mind and body.

The same applies to weight control. Getting weight under control certainly means paying attention to the body—making more healthy food choices, cutting back on the amount of food we eat, and exercising more. But none of these bodily changes can happen, or can be sustained in the long term, if our minds aren't well fed with nourishing thoughts that help us stay on track—and that address the issues that caused us to gain weight in the first place.

The four nutriments taught by the Buddha provide the path for doing just this.

When most scientists think of nutriments, they think of foods such as nuts, fruits, and vegetables; drinks such as juice or milk; and nutrients such as protein, fats, carbohydrates, vitamins, and minerals. Buddhist teaching, however, offers a more inclusive way of looking at nutriments. Besides edible food and drink—the nutrients that sustain our bodies and feed our brain—there are three other types of nutriments that enable us to preserve the health and well-being of our body and our mind. These other nutriments are the following:

SENSE IMPRESSIONS: what we see, hear, taste, smell, touch, and think

VOLITION: our inner motivations, our deepest desires

CONSCIOUSNESS: the totality of everything we have thought, said, or done throughout our lives as well as the knowledge, habits, talents, and perceptions of our ancestors. Thus consciousness is both individual and collective.

If we experience a problem in our body or a disturbance in our feelings, our mind, or our consciousness, we need to identify what types of nutriments we have been feeding ourselves that have led to our negative state. Once we have identified those nutriments, we can stop ingesting them and, in turn, heal the problem areas. For example, if we find ourselves easily getting angry, agitated, or sad, which then causes us to eat too much out of frustration, we need to look deeply to see what has brought about our anger, agitation, or sadness: What foods have we eaten? What types of sensory input have we taken in? What are the intentions that drive us, and what is the state of our consciousness, in this moment and as an accumulation of experience over the course of our life? Maybe we have read glossy magazines full of advertisements for clothes and accessories we cannot afford and do not need, and this has made us feel anxious and inadequate. Maybe we are frustrated that loved ones don't act as we wish, which fills us with anger and resentment. Once we identify what nutriments we are consuming that are harming us or others, we can work to change our actions and find healthier ways to deal with our obstacles. This will not only help with our well-being but also keep us from gulping down calories to deal with our difficult emotions.

The First Nutriment: Edible Food and Drinks

The first nutriment is essential for our well-being. What we eat and drink, and the way we eat and drink, profoundly affect our physical and mental well-being. That is why it is essential to know which foods and drinks promote health and which foods and drinks harm us. Nutrition research over the past fifty years has found that maintaining a healthy eating pattern can reduce our risk of major chronic diseases including diabetes, heart disease, obesity, and cancer. This scientific nutrition advice is summarized in chapter 5.

As modern society has learned more and more about what constitutes a healthy diet, our current food industry system has become more and more complex. We no longer grow our own food, and we rarely buy our foods from local farms offering basically whole foods with minimal processing and no pesticides. Nowadays, most of us buy foods from supermarkets that have tens of thousands of items for us to choose from.[4] Each year, the U.S. population spends roughly 10 percent of its income on food—roughly one trillion dollars in 2008 alone.[5] And each year, thousands of new foods and drinks are introduced. Many of these products are highly processed and loaded with sugar, salt, or refined carbohydrates, which compromise our health.

The choices are mind-boggling. Walking down supermarket aisles at eye level, you find many varieties of snack foods such as cookies, snack and cereal bars, chips, and drinks that have a long list of unrecognizable ingredients. Though some may make nutritional claims that they are high or low in certain nutrients and are good for you, these claims can be deceiving. For example, a cereal bar may be enriched with vitamins and minerals, but it likely contains too much sugar and refined carbohydrate to be considered a healthy choice. In this world of abundant choices that are designed to appeal to our tastes and our desire for convenience, if we are not mindful as we scan the supermarket shelves, we could end up buying and consuming foods and drinks that insidiously hurt our health without our being aware of it.

Mindfulness also helps us look beyond the packaging to see how we grow and where we get our food, so that we can eat in a way that preserves our collective well-being and the well-being of our planet. If we do not take care of our planet, we will not have the adequate sunshine, air, temperature, rain, clean water, and wholesome soil needed to grow our food. Instead we will have unwholesome and contaminated foods that harm our body, our mind, and our world. We have to know what we are eating, where our food is from, and how it affects us.

The Buddha specifically advised us to eat mindfully so that we can maintain compassion in our hearts and ensure a good future for the next generations. He taught that if we take a shortsighted and selfish approach to the food and drink we consume, we will hurt not only ourselves but also our children and our planet.

One teaching from the Buddha that addresses this issue directly is the Sutra on the Son's Flesh. This parable may sound unimaginable, cruel, and totally unacceptable. But it contains a powerful lesson about the foods we consume and the future of our planet.

Sutra on the Son's Flesh

A young couple and their three-year-old son had to cross a vast desert and move to another country, where they wanted to seek asylum. They were not familiar with the terrain, nor did they know how long the journey would take, and they ran out of food while they were only halfway through the desert. They realized that without enough food, all three of them would die in the desert, with no hope of reaching the country on the other side of the desert. After agonizing reflection, the husband and wife made the decision to kill their little son for food. Each day they ate a small morsel of his flesh in order to have enough energy to move on, and they carried the rest of their son's flesh on their shoulders so that it could continue to dry in the sun. Each time when they finished eating a morsel of their son's flesh, the couple looked at each other and asked, "Where is our beloved child now?"

Having told this tragic story, the Buddha looked at the monks and asked, "Do you think that this couple was happy to eat their son's flesh?" "No, World-Honored One. The couple suffered when they had to eat their son's flesh," the monks answered. The Buddha taught the following lesson: "Dear friends, we have to practice eating in such a way that we can maintain compassion in our hearts. We have to eat in mindfulness. If not, we may be eating the flesh of our own children."

The story may be extreme, but we need to wake up so that we are not, albeit figuratively, consuming our children's flesh and experiencing the pain of the couple. In fact, much of the world's suffering comes from not eating mindfully, from not looking deeply into what and how we eat. This mindless eating can lead to the weight we gain and the diseases caused by poor nutrition, and it takes a toll on the health of the planet also. We have to learn ways to eat that preserve the health and well-being of our body, our spirit, and our planet. (Appendix B contains the sutra in its entirety, see p. 241.)

Looking deeply at the way we eat from a global perspective, we can see that meat production is a huge drain on the planet. The United Nations report *Livestock's Long Shadow,* an in-depth assessment of the damaging impact of livestock on our environment, concluded that livestock's negative effect on our environment is massive and that we need to address it with urgency. The report estimates that raising livestock uses 8 percent of our planet's water and contributes strongly to water depletion and pollution.[6] Some scientists have estimated that it takes one hundred times as much water to produce a kilogram of beef as it does to produce a kilogram of protein from grain.[7] Part of the reason that so much water is needed to produce livestock is that cattle are fattened on vast amounts of grain, which requires water to grow. In the United States, cattle consume seven times as much grain as the U.S. population as a whole.[8] An Environmental Protection Agency report on U.S. agricultural crop production in 2000 states that, according to the National Corn Growers Association, about 80 percent of all corn grown in the United States is consumed by domestic and overseas livestock, poultry, and fish production.[9] Yet, ironically, more than nine thousand children die each day from causes related to hunger and undernutrition.[10] It is a painful realization that the grain and resources we use to raise livestock could be used more directly instead to feed the starving and malnourished children in the world.

Furthermore, a 2008 report by the Pew Charitable Trust and the Johns Hopkins Bloomberg School of Public Health found that factory farming in the United States is taking a heavy toll on human health and the health of the environment—and that keeping livestock in these "concentrated animal feeding operations" constitutes inhumane treatment.[11] Animal waste pollutes the water and air around the farms, causing illness among farmworkers and farm neighbors, as well as land degradation. Heavy use of antibiotics in factory farming leads to new strains of viruses and bacteria resistant to antibiotics, creating "superbugs" that may pose a public health threat to us all. In the report, the experts recommended phasing out and banning the use of antibiotics in farm animals except for the treatment of disease, instituting tighter regulation of factory farm waste, and phasing out intensive confinement systems.[12]

The devastating environmental and societal impact of raising livestock goes beyond the use of water and land to grow food. Our society's hunger for meat contributes mightily to the production of climate-changing greenhouse gases. The livestock industry is responsible for 18 percent of the world's greenhouse gas emissions, a higher share than the entire transportation sector.[13] Seventy percent of forests in the Amazon have been cut to provide grazing land for cattle, and when such forests are destroyed, enormous amounts of carbon dioxide stored in trees are released into the atmosphere.[14] The meat, dairy, and egg industries are also responsible for two-thirds of human-induced emissions of ammonia, which in turn plays a role in acid rain and the acidification of our ecosystem.[15]

The data suggest that one of the best ways to alleviate the stress on our environment is to consume less meat and eat more plant-based food, which results in reduction in greenhouse gas emissions. We do not need cattle to process the food for us. It is much better and more efficient for us to eat more plant-based food and process it ourselves. It may seem like a huge change for many people, but reducing the amount of meat and dairy in your diet is a great way to keep your weight in check, improve your overall health, and take steps toward improving the health of our planet. When we learn to eat more vegetables, grains, and beans mindfully, we will enjoy their taste, and we can be happy knowing that we are supporting a new kind of society in which there is enough food for everyone and no one will have to suffer from hunger.

We must take urgent action at the individual and collective levels. For individuals, going toward vegetarianism can have great weight and health benefits. Vegans and vegetarians tend to weigh less than people who consume animal products; they also tend to have lower risk of heart disease, diabetes, and some cancers.[16] In chapter 5, we go into greater detail about the health benefits of plant-based diets.

Many Buddhist traditions encourage vegetarianism. Although this practice is primarily based on the wish to nourish compassion toward animals, it also offers many health benefits. Now we also know that when we eat vegetarian, we protect the earth and help reduce the greenhouse effect that is causing her serious and irreversible damage. Even if you cannot be 100 percent vegetarian, being a part-time vegetarian and consuming a more plant-based diet is already better for your own health as well as the health of our shared planet. You may want to start by eating vegetarian for a few days a month, or you can eat vegetarian only for breakfast and lunch every day. This way, you are already more than half vegetarian. If you feel that you cannot eliminate animal products from your diet for even one meal, simply reducing the portion of meat and eliminating processed meats like bacon, sausages, and ham can lower your risk of colon cancer and your risk of dying an early death from heart disease, cancer, or other causes.[17] This is a good first step to adopting a more plant-based, healthful, environmentally friendly diet.

Using mindfulness to look deeply at what you eat can make it much easier to make such changes, because you realize the benefits they can bring to the planet and yourself—lower weight, lower risk of colon cancer and heart disease, and more energy for doing the things you enjoy. We are "interbeings": we and our environment are interdependent. And even small changes on our part can have a large impact when combined with others. Our market economy is driven primarily by consumer demand. As a population, if a large number of people make even small moves to eat less meat and more plant-based foods, the livestock industry will shrink. Over time, farmers will find other crops to support their livelihoods. Through such collective awakening we can make a difference in our world.

Second Nutriment: Sense Impressions

Sense impressions arise from the sensory activities and responses of the six sense organs, the six sense objects, and the six sense consciousnesses. The six sense organs are the eyes, ears, nose, tongue, body, and mind. The six sense objects are form, sounds, smells, tastes, tactile objects, and mental objects, or objects of mind. The six sense consciousnesses are eye consciousness (or sight), ear consciousness (or hearing), nose consciousness (or smelling), taste consciousness (or tasting), body consciousness (or touching), and mind consciousness (thinking). Objects of mind include all physiological, physical, and psychological aspects of our senses.

What we see, hear, smell, taste, touch, and think, all that we sense in our body and all that we become aware of in our mind, is food for our sense consciousnesses. Throughout our waking hours, our six sense organs are actively engaged. The nutriments we ingest through our six senses can be either healthful or harmful—especially when it comes to our attempts to reach a healthier weight. Think of a day in your life. As you get up, you turn on the radio and your favorite piece of music is playing. Because your ears are functioning well, you can hear the music, and you feel good and lighthearted. That piece of music lingers in your consciousness, and for the next few hours you find yourself humming it and smiling to yourself. During your lunchtime walk, you hear that same piece of music on your MP3 player, and it brings a lightness and energy to your step. You walk by a bus that has an ad on its side promoting a new television thriller series, and you store the information in your consciousness. After work, you stop at the supermarket, intending to buy some healthful food for dinner, and you pick up a magazine at the checkout counter. Flipping through the pages, you see an ad for fancy chocolate, and the woman who eats it looks relaxed as she enjoys this indulgence. As you wait in line, you can see and smell the chocolates that line the checkout aisle, and you decide to throw a few chocolate bars into your cart. That evening, you turn on the television to watch the thriller advertised on the bus that you noticed during your afternoon walk. As you watch the program, you become tense and edgy because there are many scary, suspenseful scenes. You then crave chocolate and decide to eat a chocolate bar before you go to bed, since somewhere in your consciousness you have the impression that chocolate can help you relax. That night, you have a vivid dream filled with suspense and fear. You wake up tense. You take another chocolate bar along with you to work to snack on at your desk during the day.

Mass media is the food for our eyes, ears, and minds. When we watch television, read a magazine, watch a film, or play a video game, we are consuming sensory impressions. Many of the images we are exposed to through the media water unwholesome seeds of craving, fear, anger, and violence in our consciousness. The images, sounds, and ideas that are toxic can rob our body and consciousness of their well-being. If you feel anxious, fearful, or depressed, it may be because you have taken in too many toxins through your senses without even knowing it. Be mindful of what you watch, read, and listen to, and protect yourself from the fear, despair, anger, craving, anxiety, or violence they promote. The material goods they promise are only quick, temporary fixes. True contentment lies within.

In the United States, consumerism dominates our culture. We are able to shop around the clock, thanks to the Internet and twenty-four-hour stores. The same is true for foods: we can buy food anywhere and at any time. And we don't really ever stop and ask ourselves: Why are we buying so much? Do we really need all this "stuff"? Why are we eating so much? Are we truly hungry?

We all really need to take a step back and take a deep look at what our true needs are. And one way to help do this is to become mindful observers of the market-driven, ad-driven world we live in. Advertising is designed to create a need where one does not exist—and it must work, because companies spend hundreds of billions of dollars a year on it.[18] Often in commercials, buying a certain food or gadget is portrayed as an antidote to loneliness or insecurity. People in commercials consuming fast food or ice cream all appear to be so happy when they are eating these products, so fulfilled and vibrant. We absorb and store these perceptions and messages in our consciousness without censoring their content. Later, we find ourselves consuming these foods, even though we know they can cause us harm. And we wonder why.

We can choose to resist these messages—but it will be easier if we mindfully choose to limit our exposure to such messages. Turn off the television. Stop mindlessly reading glossy magazines. Children especially need protection from the media, since their minds simply are not mature enough to understand that advertisers are deliberately trying to influence them.[19] We also need to shield ourselves and our children from unwholesome films, TV programs, and video games, in addition to advertisements, because they can fill us with anxiety, violence, and craving. They can also fill us with stress, and stress may, in turn, contribute to weight gain.[20] When we spend a lot of time in the sun, we can wear sunscreen to protect ourselves from harmful ultraviolet rays while still being able to enjoy the warmth of the sun. In the same way, mindfulness is the shield that can protect us from these corrosive and stressful messages in our everyday environment while helping us filter and choose positive, wholesome sense impressions that water seeds of happiness and peace in our consciousness so that we are less likely to eat out of our negative emotions.

Learning to mindfully consume sensory impressions can help us reduce our craving, anger, fear, sadness, and stress. And all of this may ultimately help us in our quest to achieve healthier weight.

Third Nutriment: Volition

The third kind of food is volition, or will—our deepest desire to obtain whatever it is that we want. What we want drives our daily actions. What we want also determines our personal aspirations. We have to ask ourselves: What is my deepest desire in this life? We have to look deeply into ourselves to see what kind of energy motivates us in our daily life. We all want to go somewhere or realize something. What is the purpose of our life? Our desire can take us in the direction of happiness or in the direction of suffering. Desire is a kind of food that nourishes us and gives us energy. If we have a healthy desire, such as a wish to save or protect life, care for our environment, or live a simple, balanced life with time to take care of ourselves and our loved ones, our desire will bring us happiness.

Everyone wants happiness, and there is a strong energy in us pushing us toward what we think will make us happy. But we may also suffer a lot because of this relentless pursuit. There are those of us who believe that happiness is possible only when we gain a lot of money, fame, and power. Yet these things can be suffering disguised as happiness, as they are often built on the suffering of others. For instance, the opium and slave trade inflicted tremendous human suffering around the world. Or a modern-day version of the slave trade: the global sex trade, which dehumanizes young women and children from many countries who are forced to work in foreign countries in the booming sex industry, sometimes with legal "entertainment visas," as in the case of Japan. The desire to make money, in and of itself, is not a bad thing if your pursuit of material wealth does not harm anyone along the way and if you use your money in a compassionate way. It's important to look deeply at our desires and see if they are built on positive or negative intentions. This can help us steer our desires toward those things that are beneficial to others, to the world, to our family, and also to ourselves.

In a 1999 meditation retreat for business leaders, many participants shared stories of how people with great wealth and power also suffer tremendously. A very wealthy businessman said that despite having over three hundred thousand employees, with operations in many parts of the world, he was extremely lonely. This man's loneliness, and the loneliness of many wealthy people, stems from being suspicious of others. They feel that those who want to befriend them do so because of their money and only want to take advantage of them. They feel lonely because they do not have any real friends. Children of wealthy people also suffer deeply; often their parents have no time for them because they are so preoccupied with maintaining their wealth and social status. The suffering of many wealthy people shows us that money cannot buy us true happiness.

Our deepest desire is the basis of all our actions, including our career. If you want to be a doctor and heal people, you will focus your energy and prepare yourself for many years, going through demanding training in medical school, internship, and residency to become a board-certified doctor. After you become a doctor, you forget about the hard work and many sleepless nights, and instead simply feel good about the contribution you are making to society. Unfortunately, there are many other professionals whose primary desire is just to make more money for themselves. The fund managers who advanced personal monetary gains for themselves during the 2008 housing bubble are now left with the realization that they had a major role to play in the crumbling of the world economy, causing many people around the world to become homeless and jobless. Can they be truly at peace with themselves, living with this realization?

We must look deeply into the nature of our volition to see whether it is pushing us in the direction of liberation from suffering and toward peace and compassion, or in the direction of affliction and misery. What is it that we really want deep in our heart? Is it money, fame, power? Or is it finding inner peace, being able to live life fully and enjoy the present moment? Happiness reveals itself when we are at peace with ourselves. We are not happy because we weigh more than we should. But weight by itself may not be the underlying cause of our unhappiness.

Desire is often at the root of weight problems: our desire to eat too much tasty food, our desire to avoid difficult emotions by distracting our minds with snacks and television, our desire to work long hours at the office for career advancement, leaving us little time to go to the gym or walk in nature. How do we balance all these desires or set priorities for them?

Taking a deep, mindful look at our true desire can help direct us on the right path to well-being. By observing the interdependent nature of our eating problems, and our volition to achieve well-being, we can certainly identify and change conditions that will bring about inner peace and joy.

Fourth Nutriment: Consciousness

Every day our thoughts, words, and actions flow into the sea of our consciousness. Our sense perceptions are continuously feeding our consciousness. The imprints of all our experience and perceptions are stored as seeds in the deepest level of our mind, called the *store consciousness.* There are also seeds that contain the inherited habit energies of all our ancestors and affect our pattern of seeing, feeling, and thinking. As long as they are in store consciousness, they are still sleeping and lying dormant. But when watered, these seeds have the capacity to manifest in our daily lives as full-grown energies. When you plant a flower seed in the spring, by summer a plant will mature and bear flowers; from these flowers new seeds are born, and the cycle continues. Similarly, seeds of compassion, joy, and hope, as well as seeds of sorrow, fear, and despair, can grow in the field of our mind. The germinating seeds grow into the upper level of the mind, called the *mind consciousness.* Mind consciousness—our everyday, waking state of consciousness—should be like a gardener, mindfully attending to the garden, the store consciousness. The gardener has only to cultivate the earth and water the seeds, and the garden will nourish the seeds to bring forth the fruit.

Our mind is the foundation of all our actions, whether they are actions of body, speech, or mind, i.e., thinking. Whatever we think, say, or do arises from our mind. What our consciousness consumes becomes the substance of our life, so we have to be very careful which nutriments we ingest. In describing the second nutriment, we talked about sensory impressions and the need to guard our senses. Our senses are sometimes referred to as *gates* because all the objects of our perception enter our consciousness through sensory contact with them. The mind consciousness, the gardener, has to be an attentive guard at these gates of our senses and carefully choose which sense impressions to allow in. Mind consciousness must also recognize and identify the wholesome seeds in store consciousness, practice day and night to care for and water those wholesome seeds and help them grow, as well as prevent any negative seeds from being watered. The way we do this is through mindfulness.

According to Buddhist psychology, when a seed rises up from store consciousness into our mind consciousness it becomes a mental formation. *Formation* is a technical term meaning something that manifests itself based on conditions—a composite of different elements or attributes joining together when the conditions are ripe. A flower is a physical formation of several elements: seed, rain, sunshine, earth, air, space, time, and so on. When these elements come together under the right conditions, a flower manifests itself.

Regarding the fourth nutriment, we are speaking not about physical formations that make up the food of our consciousness but about mental formations. Fear is a mental formation. It is composed of several mental and emotional elements: anxiety, doubts, insecurity, misperceptions, and ignorance. Despair, anger, love, and mindfulness are other examples of mental formations. These are merely symbols or names that we use to describe the experience resulting from the interactions between our sense organs and their sense objects, which give rise to all kinds of mind states, including responses to thoughts, feelings, perceptions, mental trauma, and memories.

Lying deep down at the bottom of the store consciousness are all kinds of seeds. All mental formations are buried in the form of seeds in the soil of our store consciousness and can manifest themselves on the upper level of consciousness, the mind consciousness. There are many different kinds of seeds living in store consciousness, both wholesome and unwholesome. Wholesome seeds include seeds of love, gratitude, forgiveness, generosity, happiness, and joy. Unwholesome seeds include hatred, discrimination, jealousy, anger, and craving. For example, our hatred is a mental formation. When it is not manifesting itself, we do not feel hate. However, that does not mean that the seed of hate is not in us. All of us have the seed of hate in our store consciousness. We can be very openhearted and loving and not feel any hate at all. However, if we encounter an unjust, oppressive, or humiliating situation that waters the seed of hatred in our store consciousness, hate will begin to sprout and grow into a zone of energy in our mind consciousness. Previously hate was just a seed, but once it has been watered, it develops and becomes the mental formation of hate. We then become angry and full of ill will, experiencing hateful thoughts and physical tension.

Whenever a seed manifests itself in our mind consciousness, we absorb it as food for our consciousness, the fourth nutriment. If we allow anger to come up into our mind consciousness and stay for a whole hour, for that whole hour we are eating anger. The more we eat anger, the more the seed of anger in our store consciousness grows. If you have a friend who understands you well and offers you words of comfort and kindness, the seed of loving-kindness will arise in your mind consciousness. If you are in the company of that kind friend for one hour, then during that time you are consuming a whole hour of loving-kindness. Any seed, wholesome or unwholesome, that has an opportunity to manifest itself as a mental formation at the level of the mind is strengthened at its root in store consciousness. Therefore, we must learn to nurture wholesome seeds and to tame unwholesome ones with mindfulness, because when they return to the store consciousness, they become stronger regardless of their nature.

We can take care not to water our unwholesome seeds—such as anger, despair, and hopelessness—by being mindful of situations that can elicit them. These situations may be from images we see in the mass media or conversations we hear—either in our interactions with others or over the airwaves. Furthermore, we can help each other water the wholesome seeds in our store consciousness by being kind, thoughtful, and understanding of others. When we water seeds of forgiveness, acceptance, and happiness in the people we love, we are giving them very healthy food for their consciousness. But if we constantly water the seeds of hatred, craving, and anger in our loved ones, we are poisoning them.

Only by looking deeply into the nature of our suffering can we discover its causes and identify the nutriments that have brought it into being. After we have practiced for some time, we will see that transformation always takes place in the depths of our consciousness; our store consciousness is the support, the base for our consciousness. If we know how to acknowledge and recognize the presence of the mental formation, embrace it, calm it, and look deeply into it, we will gain insight. This insight can liberate us and transform our afflictions in their form as seeds, so that they no longer arise in mind consciousness.

How does this relate to our difficulty with weight? We must find the source of our desire to eat too much of the wrong foods. Perhaps we eat out of sadness; perhaps we eat out of our fears for the future. If we cut the sources of nutriment for our sadness and fear, sadness and fear will wither and weaken, and with them the urge to overeat. The Buddha said that if we know how to look deeply into our suffering and recognize its source of food, we are already on the path of emancipation. The way out of our suffering is through mindfulness of consumption—all forms of consumption, and not just edible foods and drinks.

When fear, despair, anger, or pain is active in our consciousness, we can draw on mindfulness to bring us relief. If anger, fear, and despair are dormant, they will not be perceptible in our consciousness, and our life will be much more pleasant. Yet we ingest the toxins of violence, fear, and anger every day from our environment, including the media. We also ingest unwholesome interactions with others or painful memories from the past. So the negative seeds are frequently watered and become stronger and stronger. These negative emotions of anger, fear, and violence then become an integral part of our everyday lives, blinding us from seeing things clearly and keeping us in ignorance, which is the cause of suffering. However, if by cutting off their food or nutriments we do not allow these seeds of negative emotions to grow, we will not be overcome by violence, fear, or anger. And we will not be driven to overeat. (See figure 3.1.)

In our store consciousness, we also have the seed of mindfulness. If we water the seed of mindfulness often, it will grow stronger. Because all the seeds are interdependent in nature—the state of one seed can influence the state of all others—a strong mindfulness energy can help us transform our negative emotions. This mindfulness energy is like a torch helping us to see clearly the true nature of our suffering. It also provides energy to manifest our seeds of wisdom, forgiveness, and compassion so that ultimately we can free ourselves from our suffering. Without wisdom, forgiveness, and compassion, happiness and peace will not be possible. Suppose we are standing in front of the refrigerator after encountering a splash of anger from a family member. We are not hungry, because it is only an hour or so after dinner. We have a choice. Either we can be totally consumed by the unpleasant incident, get distressed, and then assuage our hurt feelings with food from the refrigerator, or we can mindfully tender the unpleasant emotions and recognize that overeating would lead us to feel even worse later—ashamed for abandoning once again our commitment to eat more mindfully—and would not help us resolve the hurt feelings from our quarrel with our family member. Mindfulness helps us free ourselves from dwelling on the unpleasant incident and stops any thoughts of revenge or overeating as it waters the seeds of wisdom and compassion in us. When we pause with mindfulness, we recognize that our family member must be suffering somehow. If one is happy and peaceful, one would not behave with such anger. Mindfulness practice can help reveal this kind of insight, which can free us from the imprisonment of past events so that we can make clear choices to help us manage our weight.

Figure 3.1 SEEDS OF MINDFULNESS

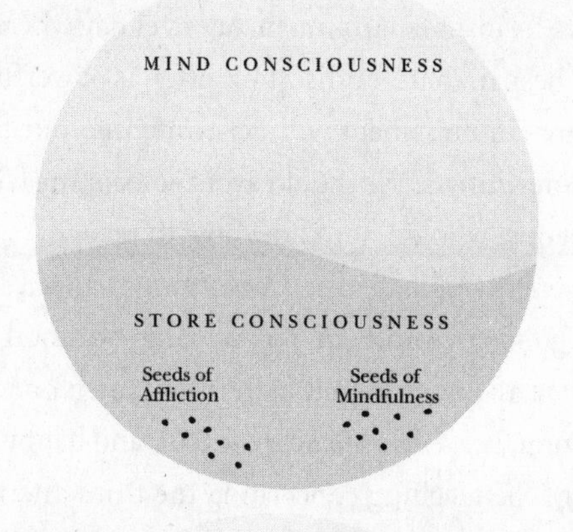

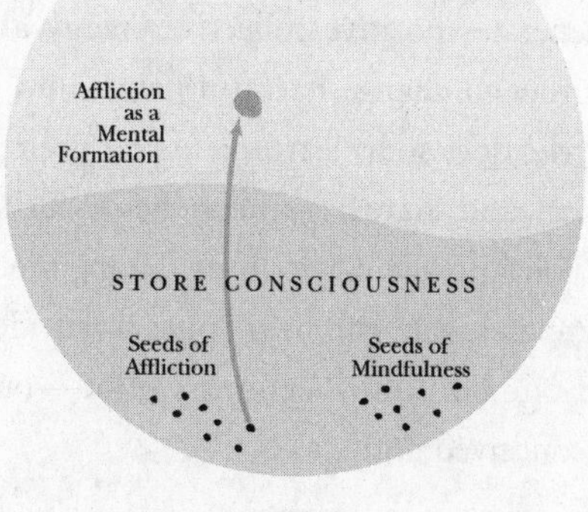

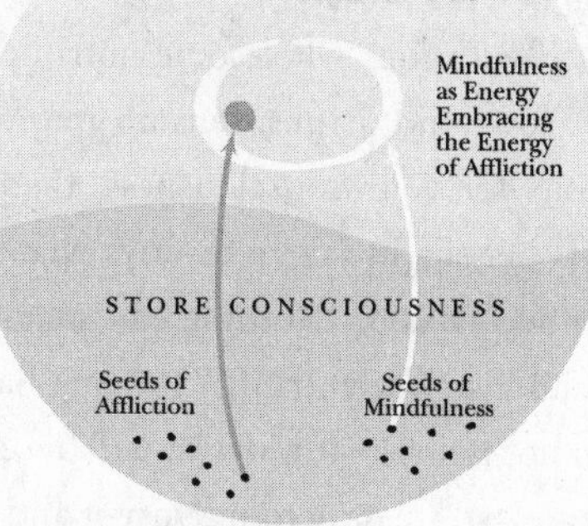

Collective consciousness is also a powerful source of nutriment. If we allow ourselves to be in an environment day after day where the collective energy of anger, despair, hate, or discrimination is powerful, then sooner or later this source of nutriment will penetrate into our body and consciousness and imprison us. We should avoid associating with individuals and groups of people who do not know how to recognize, embrace, and transform their energy of hate, discrimination, or anger. It is important that we select a good environment, a good neighborhood for us and for our children. Such an environment will help nourish our ideals and our wholesome volition, keeping us healthy, joyful, and happy.

In the light of the teaching concerning the third and fourth sources of nutriment, we see that it's beneficial to look for and to live with people who share the same ideal, intention, and purpose. This is a *sangha,* a community that generates positive collective energy and is motivated to support one another to change harmful habits into beneficial ones. Everyone in the *sangha* does so by learning to practice restraint, observe the law of moderation, and share happiness, thereby bringing a spiritual and ethical dimension into their daily lives. Living among people who are healthy and compassionate will help nourish our ideals, our wholesome volition, and our beautiful beginner's mind—our ability to see things without preconceived notions.

Nourish Yourself Mindfully with Four Nutriments Every Day

If you nourish yourself with four wholesome nutriments, consuming a healthy diet of edible food and drinks, sensory impressions, intention, and mental formations for your consciousness, then you, along with your loved ones, will benefit in concrete ways that are noticeable in your daily life. The Buddha said, "Nothing can survive without food." This is a very simple and very deep truth. Love and hate are both living phenomena. If we do not nourish our love, it will die and may turn into hate. If we want love to last, we have to nurture it and give it food every day. Hate is the same; if we don't feed it, it cannot survive.

Nourishing your body and mind with wholesome nutriments will help you achieve peace and happiness and bring you further along the path of healthy weight. And it is important to realize that the mind and body are not separate. To achieve our goal of a healthier weight, we need to consume *all* of the four nutriments mindfully. We cannot just focus on one aspect of our being as if it is an entity separate from the rest. We need to address all aspects simultaneously, as a whole. Your previous weight loss efforts may have failed because of the lack of this holistic approach. You now know the essential elements needed to help you establish healthy, wholesome habits for life. This is a worthwhile journey to embark on. It will lead you in the direction of emancipation from all suffering and afflictions—to the heart of the issue and the underlying root of your unhealthy weight. When practiced consistently, each step of this mindful journey can astonish you, bringing you greater realization of confidence, joy, and peace.

CHAPTER

Stop and Look: The Present Moment

TO UNDERSTAND AND TRANSFORM our suffering, we need to use a step-by-step process of deep observation—just like the scientist does in the laboratory. We start by being aware of and acknowledging our suffering. Take, for example, our unhappiness with our weight. We must make an effort to stop our busy lives for a moment and become aware of our suffering—something most of us try to avoid and resist. Instead, we need to embrace and accept the pain we feel about our weight. Next, we must realize that the suffering inside us isn't just something we are looking at from the outside: We are that block of suffering. We become one with our suffering, just as the scientific observer becomes one with the objects of his investigation, and this amalgam is the key to transforming and relieving our misery. For example, to understand our shame of being overweight, we must acknowledge and accept that we are embarrassed, angry, and filled with despair. By becoming one with our suffering, we can feel it. We acknowledge the feelings; we do not reject them or push them away. We know that we can identify the causes of our suffering, and we can find a way out. How can we become deep observers of our suffering and free ourselves from it? Through the daily practice of mindfulness.

So What Is Mindfulness?

In the previous chapters, we have briefly mentioned mindfulness. Now we are going to offer a deeper, more comprehensive understanding of mindfulness. The Chinese character for "mindfulness" is "nian" (念). It is a combination of two separate characters, each with its own meaning. The top part of the character (今) means "now," and the bottom part of the character (心) means "heart" or "mind." Literally, the combined character means the act of experiencing the present moment with your heart. So mindfulness is the moment-to-moment awareness of what is occurring in and around us. It helps us to be in touch with the wonders of life, which are here and now. Our heart opens and is immersed in the present moment, so that we can understand its true nature. By being present and mindful of the present moment, we can accept whatever is at that moment as it is, allowing change to happen naturally, without struggle, without the usual resistance and judgment that cause us to suffer more.

Mindfulness is the energy that helps us look deeply at our body, feelings, mind perceptions, and all that is around us. It is a source of light in the darkness, allowing us to see clearly our life experience in relation to everything else. It is through this kind of insight that we can lift ourselves out of ignorance, the main cause of suffering.

Though many books have been written about the power of mindfulness, it is something best learned by doing. Just like a toddler learning to walk through repeated effort—learning to crawl, then standing up, falling countless times, and standing up again—we must, if we want to get a good handle on mindfulness, constantly apply it to everything we do, so that it becomes second nature. Mindfulness does not happen by itself, however. You need to have the desire to practice it.

Mindfulness can help us understand ourselves in relation to all that is within and around us. When we have difficulty with our weight problems, we often resent ourselves. We tend to look at our eating habits and our distress as separate entities from ourselves, and try to solve these problems from the outside. We need to compassionately realize that these problems are not separate from us: they are our own body, feelings, and mind, which are interconnected with everything else in our world. This deep understanding of the interdependence of all things enables us to recognize what can be done to effect change in a steady manner.

In the Discourse on the Four Foundations of Mindfulness, the Buddha taught his disciples how to practice mindfulness to "help living beings realize purification, overcome directly grief and sorrow, end pain and anxiety, travel the right path, and realize nirvana." Although the Buddha was not specifically addressing weight management, the basic guidance is most relevant. The insight it offers is as applicable now as it was then.

The Four Foundations of Mindfulness

There are four foundations of mindfulness. The first is *our own body.* When dealing with weight and eating problems, naturally, we need to first know our bodies and how to be in touch with and appreciate them. The second foundation of mindfulness is *feelings.* It teaches us how to be aware of our physiological, physical, and psychological feelings in broad categories of pleasant, unpleasant, mixed and neutral feelings. The third foundation is *mental formations*—such as compassion, anger, or greed—the more complex reactions that arise based on our feelings. This foundation involves the practice of being aware of the mind's activities. The fourth foundation is the realm of *objects of mind,* because each mental formation has an object. If there is no object there is no subject, because consciousness is always consciousness of something. The fourth foundation is the awareness of all things in us and around us, the objects of our mental formations. Central to each area of practice is the understanding that we are not separate from the object of our mindfulness. As we watch our bodily sensations or emotions, we are feeling them at the same time. Furthermore, although each area of mindfulness focuses on a different object of observation, all four areas are interconnected.

Mindfulness of the Body (Observation of the Body in the Body)

Mindfulness of the body is simply that: observing and becoming one with our body and its condition. Being mindful of the body, we practice observing and becoming fully aware of the breath, the positions of the body, the actions of the body, and the various parts of the body. We become aware of the state of the body, including our aches, our pain, and our overweight. This is an important practice because in our busy lives, we often ignore warning signals from our body and postpone responding to its cries for help until it is too late.

The first important practice is the full awareness of breathing. Try sitting in a chair comfortably, with your feet firmly touching the floor and your back straight. If you prefer, you can also lie down flat and relax your body. Shift your attention to your in-breath and out-breath. Say silently:

When I breathe in, I know that I am breathing in.
When I breathe out, I know that I am breathing out.

This conscious breathing exercise is simple, yet its effects are profound when practiced regularly. To succeed, we must give all of our attention to our breathing, and nothing else. As we follow our in-breath, for example, we feel the air flowing into our nostrils and our lungs.

When distracting thoughts arise, we let them go and refocus on our in-breath and out-breath. Our mind stays focused on our breathing for the entire length of each breath. As we breathe, we become one with our breath. The body and the breath are not separate entities. This is "mindfulness of the body *in* the body."

In everyday life, we often get lost in forgetfulness, operating on autopilot for most of our waking hours. Our mind chases after thousands of things, and we rarely take the time to come back to ourselves, to be in touch with ourselves. We end up feeling overwhelmed and alienated from ourselves. Conscious breathing is a marvelous way to return to ourselves, like a child returning home after a long journey. When we are still, we feel the calmness that we have inside us, and we find ourselves again. Conscious breathing also allows us to be in contact with life in the present moment, the only moment in which we can truly touch life.

When we follow our breathing, we are already at ease, no longer dominated by our anxieties, resentments, and longings. As we breathe consciously, we become more stable with every moment.

Take a moment to try this simple breathing exercise, and observe how you feel afterward.

When breathing in a long breath, silently say to yourself, "I am breathing in a long breath." Then repeat the word "long" with each in-breath.

When breathing out a long breath, silently say to yourself, "I am breathing out a long breath." Then repeat "long" with each out-breath.

Or if your breath is short, when breathing in, say, "I am breathing in a short breath." Then repeat "short" with each in-breath.

When breathing out a short breath, say, "I am breathing out a short breath." Then repeat "short" with each out-breath.

While the mind is following the breath, the mind is the breath and only the breath. The two unite and influence each other. In the process of the practice, our breathing naturally becomes more regular, harmonious, and calm, and our mind also becomes more regular, harmonious, and calm. When the mind and the breathing unite, feelings of joy, peace, and ease arise in the body.

Breathing in, I am aware of my whole body.
Breathing out, I am aware of my whole body.

With this exercise, the distinction between body and mind dissolves, and we experience the oneness of body and mind. The object of our mindfulness here is no longer simply the breath but the whole body itself as it is unified with the breath. In our daily lives, our mind and our body do not often work in unison. Our body may be here while our mind is somewhere else, perhaps regretting the past or worrying about the future. And this disconnect between mind and body is the crux of many weight problems. For example, many people eat without feeling hunger or eat beyond the point of fullness, either because the food looks good to them and they crave it or because they are trying to soothe their difficult emotions. Through mindfulness practice, we can nurture the oneness of body and mind, and really listen to our body and know what it needs to be truly nourished. We are able to restore our wholeness and eat what our stomach wants, not what our craving pushes us to eat.

As body and mind become one, we need only calm our body in order to calm our mind.

Breathing in, I calm my body.
Breathing out, I calm my body.

The essence of mindfulness is to come back to dwell in the present moment and observe what is happening. When body and mind are one, the wounds in our hearts, minds, and bodies begin to heal. Then we can truly begin to transform our weight issues.

We have all had bad days when everything seems to go wrong. After such a long day, we feel tired, discouraged, and down. We might have an urge to have some comfort food—a pint of ice cream, say, or some chocolate-chip cookies or a bag of chips. At times like these, it is best simply to return to our body through mindful breathing, cut off all external contact, and close the door of the senses. Following our breathing, we can collect our mind, body, and breath, and they will become one. We will feel warm and soothed, like someone sitting indoors by the fire while the wind and rain are raging outside.

This method can be practiced anywhere at any time—while waiting in line, on the train, on the plane, and in our office. We can practice the same breathing technique when we walk, sit, stand, eat, drink, cook, or play. We can use our breathing in order to be aware of the positions of our body: lying, sitting, standing, or walking. We can say, "Breathing in, I calm my body," in order to continue dwelling in mindfulness and to calm our body and mind. We can come back and make ourselves whole again whenever we want or need to do so.

Practicing breathing while reciting verses like the ones described earlier in this chapter helps us dwell more easily in mindfulness. Mindfulness makes every action of our body more serene, and we become the master of our body and mind. Without mindfulness, our actions can be hurried, imprudent, insensitive, and abrupt. Mindfulness nurtures the power of concentration in us. As we practice, we find that our actions slow down. We will see our everyday actions become harmonious, graceful, and measured. Mindfulness becomes visible in our actions and speech. We are more in the flow of life, and actually living it. When any action is placed in the light of mindfulness, the body and mind become relaxed, peaceful, and joyful.

Going a little deeper, in a sitting or lying position, conscious breathing can also help us to be in touch with the functioning of our body. We observe all the parts of our body from the top of our head to the bottom of our feet. In the process of our observation, we visualize and get to know each part of our body, including the brain, heart, lungs, liver, stomach, gallbladder, spleen, blood, immune system, kidneys, bones, and so on. For example:

> Breathing in, I am aware of my liver.
> Breathing out, I know that my liver is working hard day in and day out to support me.

Because of excess weight, some people may have very negative feelings about their bodies. But if they stop and reflect, they can appreciate their eyes, feet, and hands, which are still functioning well even though other parts such as their joints and heart need more tender loving care. Many of us have lost touch with our body. Our body may have been calling out for help, but we are so preoccupied with our life's demands that we ignore our body's SOS signals. For years and years, our eyes, feet, heart, lungs, and other body parts have devotedly and faithfully worked for us nonstop. If our body is not happy and is in turmoil, we cannot be happy. Yet we rarely give the parts of our body much attention or express our gratitude to them. We cannot take our body for granted. Observing our body mindfully offers us a chance to thank our body for its hard work, enabling us to do so much in our lives, manifesting our life's work. Our body needs our appreciation and caring. The regular practice of total relaxation (see appendix C), in which we lie down and consecutively relax each part of our body, is very important for our well-being. Make a habit of practicing it regularly, starting with once a week and increasing the frequency as you experience its benefits.

Mindfully observing the different parts of the body can open the door to healing. At first we recognize the presence of the body part being observed, and then we embrace it lovingly. We see that each organ is dependent on the function of all other organs and that every muscle and cell in our body is supporting these organs.

Another exercise proposed by the Buddha in his teachings on mindfulness of the body is to see the four elements in the body: earth, water, fire, and air. You may like to see a cloud in your body, because without clouds there can be no rain, nor water to drink or grains and vegetables to eat. We see earth in us, earth as the minerals in our body. We also see the earth is in us because, thanks to Mother Earth, we have food to eat. We see air in us, because without air we could not survive, just like every other species on earth. The fire in us is the warmth made by the burning of energy from food, and it is a reflection of the sun, the fire element outside us. Everything is interconnected. Our body and our universe are one.

This concept, what we call "interbeing," applies to everything. Look into your body. Your body cannot exist alone, by itself. It has to "inter-be" with the earth, the air, the rain, the plants, your parents, and your ancestors. There is nothing in the universe that is not present in your body. When you touch your body deeply, you touch the whole universe.

One of the more difficult mindfulness exercises offered by the Buddha to help us be mindful of the body in the body is to observe our body in the different stages of disintegration after we die. Although it may be unpleasant to visualize death, and the process by which our physical body reduces to dust, the effect of this practice can be very transforming. The intention is not to make us weary of life but to provide insight into how precious life is. Many of us tend to think we will live forever, or at least that death can be put off for a long time. Many survivors of cancer, heart attack, or natural disasters such as tsunamis and earthquakes have a very different perspective on life after those traumatic experiences. Having been close to losing their lives, many of them find themselves appreciating life much more. They savor each moment and become more appreciative of the many things that they previously took for granted. But we do not have to experience this kind of trauma to see the impermanent nature of life or enjoy it fully.

Breathing in, I am aware of the impermanent nature of my body.
Breathing out, I smile to my body and enjoy being alive.

Contemplating just the body alone can already lead us to enlightenment, because the body contains the other three foundations of mindfulness as well as the whole cosmos. When we see all the conditions that come together to make the body manifest, we don't underestimate it or take it lightly. Contemplating the body is the same as contemplating the Buddha.

Sometimes we despise our body and criticize it, but even the negative things in our body are wonders. Just as with a lotus pond, we tend to accept the lotus but not the mud in the pond. The lotus is precious, but so is the mud, because without the mud there would be no lotus. We must treasure everything that belongs to the body and not despise it. It is our very vehicle for awakening. We can find all aspects of the path to enlightenment in our own body. We can't find enlightenment or nirvana outside our body; it can be found only in our body.

To be in touch with all the different aspects of the body is a main goal of the meditative, mindful observation of the body. We cannot function well without a healthy body. With the mindfulness exercises given here, we can maintain health in our body, mind, and spirit. By doing so, we free ourselves from suffering, leading to joy and happiness. When we nourish our body with joy and happiness, we can heal the wounds within us—the wounds that often prevent us from following a lifestyle that allows us to reach our healthy weight.

Mindfulness of the Feelings (Observation of Feelings in the Feelings)

Many of us approach weight loss with the desire to change only what we do not like about ourselves. But taking time to increase our joy and the wholesome qualities in us can also help us in reaching a healthy weight.

It is very important to be aware of not only what is not going well in us, but also what is going well. In every moment, there are many things that we can appreciate and enjoy, things that nourish our happiness. There are the blue sky, the white clouds, the sunshine, the solid earth beneath our feet. There are the birds singing, the trees, the presence of our loved ones, the fact that we still have food to eat, and the fact that we are still alive. Life is a miracle, and being aware of simply this can already make us very happy.

It is a very helpful practice to write down all our current conditions for happiness. We can list all the many things we have that we are grateful for and that we usually take for granted. When you brush your teeth, you can be happy that you still have teeth to brush! When you go to the toilet, you can be happy that you can still urinate and defecate on your own; you haven't lost control of your eliminating functions. You still have the use of your eyes, ears, body, and mind. When we stop to think about all these things, our list can become quite long! Observation of the feelings involves consciously bringing up our positive feelings. In the Buddha's Full Awareness of Breathing teaching, the first thing he proposes when it comes to awareness of the feelings is to nourish our joy and happiness. In order to have the strength and energy to embrace painful feelings, we must nourish our positive feelings regularly. This same practice appears in modern medicine. Before performing surgery, doctors assess whether the patient is strong enough to undergo the procedure. If not, they help the person become stronger before they operate.

Weight is, for many people, an emotional issue. Some people eat in response to emotions—happiness, sadness, anxiety, even boredom. For others, excess weight may cause them emotional strain due to the stigma and prejudice they face because of their size and appearance. Mindfulness can help us cope with these emotions and feelings.

In Buddhism, there are four sorts of feelings—pleasant, unpleasant, mixed, and neutral. Mixed feelings are both pleasant and unpleasant at the same time. Neutral feelings are neither pleasant nor unpleasant. All four are important, and none should be brushed away. When we come across an unpleasant feeling, we should not bury it at the back of our mind. Instead, we need to breathe consciously and observe it.

Breathing in, I know that this unpleasant feeling has arisen within me.
Breathing out, I know that this unpleasant feeling is present in me.

Whenever there is a pleasant, unpleasant, mixed, or neutral feeling, we need to practice mindful observation of that feeling. We need to acknowledge it and know that there is no separation between us and our feeling. We are neither drowned in nor annoyed by the feeling, nor do we reject it. This is the most effective way to be in contact with feelings. Our attitude of not clinging to or rejecting our feelings helps us avoid intensifying the feeling, and we begin our path of transformation.

Our feelings usually play an important role in directing our thoughts and our mind. When we are mindful of our feelings, the situation begins to change. We are our feelings, but we are also more than just our feelings. When mindfulness arises, there is an energy that can embrace our feeling, and then the feeling is no longer the only thing present in us and can be transformed under the light of our awareness. With mindfulness, we will no longer be swept away by the river of our feelings. When you feel anxious, instead of reaching into the freezer for an ice-cream bar, pause and consciously take a few in-breaths and out-breaths, embracing your feeling with your breathing.

Breathing in, I am feeling anxious.
Breathing out, I cuddle my anxiety.

When we accept our anxiety with compassion, we are more able to see the nature of our anxiety and more capable of transcending it. We will not be led astray by our anxiety and our habit energy of eating an ice-cream bar when we are anxious. Our mindfulness energy leads us to understand our feeling and give ourselves what we truly need to calm our anxiety.

All our feelings have a physiological or psychological root. For example, if you have an unpleasant feeling of irritability because you overate and now have indigestion, your unpleasant feeling has a physiological root. If you have an unpleasant feeling of frustration because you cannot fit into the jeans that you bought last year, your unpleasant feeling has a psychological root. To be able to identify the roots of your feelings is to look deeply in order to see how and why your irritability or frustration arose and to understand its true nature. What past experiences caused you to feel vulnerable and irritated easily? To know a feeling is not just to see its roots, but also to see its flowering and its fruits, and what it has grown into. Looking deeply, you realize that the reason you cannot fit into those jeans is that you stopped exercising because your new job is so demanding that you do not have time to exercise. Further, you recall that you always felt better and handled stress more aptly with regular exercise.

When afflictions such as anger, confusion, jealousy, and anxiety arise in us, they generally disturb our body and mind. We lose our peace, joy, and calm, and many of us turn to food, the TV, or the Internet for relief. Yet to truly regain our peace, joy, and calm, we must again practice breathing mindfully.

Breathing in, I know I have an unpleasant feeling.
Breathing out, I am here for this unpleasant feeling.
Breathing in, I calm the feeling in me.
Breathing out, I calm the feeling in me.

We have to face our unpleasant feelings with care, affection, and nonviolence. We should learn to treat our unpleasant feelings as friends who can teach us a great deal. Just like a mindfulness bell, unpleasant feelings draw our attention to issues and situations in our lives that are not working and that need our care. Proceeding with mindful observation, we will gain insight and understanding into what needs to be changed and how to change it.

With this practice, our breathing becomes lighter and calmer. As a result, our mind and body will slowly become light, calm, and clear as well. Every time we see the substance, roots, and effects of our feelings, we are no longer under the control of them. The whole character of our feelings can be changed just by the presence of mindfulness energy.

We must also practice embracing our neutral feelings. If left unattended, neutral feelings may slowly turn into unpleasant feelings. However, if we know how to manage them and if we hold them in mindfulness, they can become pleasant feelings. For example, after gaining weight, you may have had the unpleasant feeling of having blood-sugar levels higher than normal, a step on the path toward diabetes. If you lose some weight, your blood sugar may return to normal, and you will feel very happy. As time goes by you may no longer feel the intensity of the pleasant feeling associated with normal blood sugar, and it may drift into a neutral feeling as you take it for granted. However, when we are mindfully aware, our normal blood sugar can be a source of pleasant feelings again.

Mindfulness of the Mind (Observation of Mind in the Mind)

The mind is a powerful entity. It can help us achieve things we once thought were impossible. It can also, if we allow it, hinder us in reaching our goals. When it comes to our weight problem, the mind can interfere with our success.

As discussed in the previous chapter, the contents of the mind are the psychological phenomena called mental formations or mental states, which are the manifestation of seeds from our store consciousness.

There are positive and wholesome mental formations, like mindfulness, compassion, and nonviolence. And there are unwholesome mental formations, like anger, hatred, and confusion. There are also mental formations that can be wholesome or unwholesome depending on the circumstances, like regret, which is beneficial when it wakes us up and helps us not repeat our mistakes, but unbeneficial when it is prolonged and paralyzing. Every time a mental formation manifests itself, you should be able to recognize it and call it by its true name. Just like when you identify herbs for cooking by their smell and appearance.

How we feel, perceive, and act depends on how our mind reacts to and interprets the interactions between the sense organs and the sense objects. For example, as you drive into work on a sunny morning, your mind may bounce from one thought to another: "My neck hurts and I feel uncomfortable [*restlessness*]. . . . The sun is warm and bright [*joy*]. . . . I am late for my meeting [*anxiety*]. . . . Why did that car cut me off [*anger*]?!"

The restlessness, joy, anxiety, and anger described here are examples of mental formations that arise from seeds that are buried in the deepest level of our mind, the store consciousness. Our store consciousness is like a field in which every kind of seed is sown—seeds of compassion, joy, hope, and mindfulness as well as seeds of sorrow, fear, and despair. Every day our thoughts, words, and deeds reinforce certain seeds in the store consciousness. When these seeds germinate, what they generate becomes the substance of our life. Our body, our mind, and our world are all manifestations of the seeds that we have chosen to water in store consciousness.

Any seed that has an opportunity to manifest at the level of mind consciousness becomes stronger. For example, if we get into an argument, the seed of anger will manifest itself in the mind as the energy of anger. If we do not take care of this energy of anger and learn how to tame it with mindfulness, when our anger calms down and returns to the store consciousness in the form of a seed, this seed will become stronger and will arise more easily and intensely the next time we encounter a frustrating situation. As with everything, all the seeds are interdependent. The manifestation of any one seed will influence all the other seeds. Mindfulness is actually one of these seeds, and if we water it often, it will also become stronger. This is why we want to continuously cultivate our seed of mindfulness—so that it gets stronger all the time and is able to shine light on all that we experience.

Taking time to observe the mind can help you become aware of mental formations that prevent you from reaching a healthy weight, and turn them into positive mental formations. The process is the same as observing the body and the feelings. We mindfully observe the arising, presence, and disappearance of the mental formations. We recognize them and look deeply into them in order to see their substance, their roots in the past, and their possible fruits in the future, using conscious breathing while we observe them. When we take the time to do this—to fully expose the mental formations to mindfulness—they naturally transform in a wholesome direction.

For example, desire means to be caught in unwholesome longing. In relation to issues with weight, the longing can be for too much food or for watching unhealthy amounts of television. Whenever our mind and thoughts turn to such desires, we first need to acknowledge them: "My mind wants me to eat more than I should." "My mind wants me to sit and watch TV rather than go for a walk." Acknowledge their existence instead of fighting, resisting, or suppressing them. By doing so, the desires will lose their strength and grip on you.

When such unhealthy desires are not present, we also need to observe that. We can practice like this: "At this time, the mind that wants to eat too much is not arising." Desirelessness, the absence of longing for something, is one of the wholesome mental formations. It gives rise to feelings of joy, freedom, peace, and ease. It is the basis of true happiness, because in true happiness there must be the element of peace, joy, and ease.

Handling Our Anger

It is very important to learn to observe our negative emotions. One common negative emotion is anger—a complex formation that is at the heart of so many people's struggles with weight, relationships, and life in general. Identifying the presence and the absence of anger in us brings many benefits. Anger is like a flame blazing up and consuming our self-control, making us think, say, and do things that we will probably regret later. For example, when we get angry about our spouse's control over what we can eat and what we cannot eat, we may say nasty things to our spouse. Later, we regret saying things that hurt our spouse. When we observe that anger is present and we mindfully identify our anger, it will lose its destructive nature. Only when we are angry and not observing our anger mindfully does our anger become destructive. When anger is born in us, we should follow our breathing closely while we identify and mindfully observe our anger. When we do that, mindfulness has already been born in us, and anger can no longer monopolize our consciousness. Awareness stands alongside the anger: "Breathing in, I know that I am angry." This awareness is a companion for the anger. Our mindful observation is not to suppress or drive out our anger, but just to look after it. This is a very important principle. Mindful observation is like a lamp that gives light. It is not a judge. It throws light on our anger and looks after it in an affectionate and caring way without judgment, like an older sister looking after and comforting her younger sibling.

When we are angry, our anger is our very self. To suppress or chase away our anger is to suppress or chase away ourselves. When anger is born, we can be aware that anger is an energy in us, and we can change that energy into another kind of energy. If we want to transform it, first we have to know how to accept it.

We can also turn our anger into something wholesome, just as we create nourishing compost for our gardens from our food scraps. If we know how to accept our anger, to stop resisting or fighting it, we can start to feel a glimpse of peace and joy. Gradually, we can transform anger completely into the energy of understanding and compassion, allowing us to love, take care of, and honor ourselves better.

As we follow our breathing and hold our anger with mindfulness, the situation becomes less and less contentious. Although the anger is still there, it gradually loses it strength because we begin to understand it and to understand better the suffering of the person who triggered our anger. With this understanding, we can forgive and let go. We can accept our anger and make peace with it.

Looking into our anger, we can see its roots, such as the misunderstanding of ourselves and others, our pain, the violence and unkindness in our society, and hidden resentment over many generations. These roots can be present both within us and in the person who played the principal role in triggering our anger. When anger arises, we first need to come back to our conscious breathing and care for our anger with mindfulness. We concentrate on our breathing in order to maintain mindfulness.

Breathing in, I know that I am angry.
Breathing out, I know that I must take care of my anger.

Breathing in, I know that anger is still here.
Breathing out, I know that anger is in me and I know that mindfulness is in me also.

Breathing in, I know that anger is an unpleasant feeling.
Breathing out, I know that this feeling has been born and will die.

Breathing in, I know that I can take care of my anger.
Breathing out, I calm my anger.

—adapted from *Transformation and Healing* by Thich Nhat Hanh[1]

Mindfulness embraces the feeling, as a mother holds her crying child in her arms and transmits all her affection and care. If a mother puts all her heart and mind into caring for the baby, the baby will feel the mother's gentleness and will calm down. In the same way, we can calm the functioning of our mind. In order to realize the state of non-anger in our conscious and subconscious mind, we have to practice meditating on love and compassion. Anger foments many of the issues we have with weight; love and compassion for others and ourselves help us better address these issues.

Loving-kindness meditation helps us develop the mind of love and compassion and is a good antidote for our anger. Loving-kindness is the ability to bring peace and happiness to ourselves and others. Compassion is the capacity to remove the suffering in us and others. The core of love and compassion is understanding—the ability to recognize the suffering in ourselves and others. We have to be in touch with the physical and psychological suffering in ourselves and others. When we are deeply in contact with it, grounded firmly in mindfulness, a feeling of compassion is born in us immediately. Because understanding is the very foundation of love and compassion, our words and actions will then become ones that reduce the suffering in ourselves and others, dissolve our resentment, and bring more happiness to us and to others at the same time.

Love Meditation

We begin practicing the love meditation below by focusing on ourselves: "I." Until we are able to love and take care of ourselves, we cannot be of much help to others. Next, we can offer the practice for others (substituting "he/she" or "they"), starting with someone we love; next, someone we like; then, someone neutral to us; and finally, someone who has made us suffer.

May I be peaceful, happy, and light in body and spirit.
May I be safe and free from injury.
May I be free from anger, afflictions, fear, and anxiety.

May I learn to look at myself with the eyes of understanding and love.
May I be able to recognize and touch the seeds of joy and happiness in myself.
May I learn to identify and see the sources of anger, craving, and delusion in myself.
May I know how to nourish the seeds of joy in myself every day.
May I be able to live fresh, solid, and free.
May I be free from attachment and aversion, but not be indifferent.

—Thich Nhat Hanh, *Teachings on Love*

If we know how to acknowledge and recognize the presence of every mental formation, embrace it, calm it, and look deeply into it, we will gain insight. When we practice observing the mind, we can better understand why it is we feel and act the way we do. Why do we make poor health decisions? Why do we surround ourselves with people who may not be good for us? Closely observing the interrelationship of all that is in and around us helps guide us on the path to understanding and overcoming our issues with our weight.

Mindfulness of the Objects of Mind (Observation of Objects of Mind in the Objects of Mind)

Observation of the *objects of mind* includes all that can be perceived of as existing in forms, feelings, and thoughts, as well as all phenomena. When we speak of "mindfulness," we must specify: Mindfulness of what? Mindfulness of breathing? Mindfulness of walking? Mindfulness of anger? You have to be mindful of something. If that something is not there, there is no mindfulness possible. So when we speak of observing the mind in the mind, we are speaking of the subject of cognition, the subject of our mental formations, the subject of mindfulness, of hate, of love, or of jealousy. And with every subject there must be an object. To love means to love what? To love whom? To hate means to hate what? To hate whom? These are what is meant by the objects of mind.

Observing the objects of mind means to see that no phenomena has an independent existence, but arises because of the numerous conditions that have brought it to be. As the Buddha said, "This is, because that is." When we look deeply into the objects of mind, we see their arising, their duration, and their fading away. But because we see they are interrelated with everything else and have no separate self, we also see that they have no birth and no death. They were already there in the conditions that created them before they were born and they continue to exist in these conditions after they fade away. When we see the source and true nature of the objects of mind that we are contemplating, they no longer bind us.

Untying Our Internal Knots

In teaching on the observation of the objects of mind, the Buddha emphasized awareness of our internal formations, mental knots or fetters, born from our habit energies and misperceptions of reality. When we immerse ourselves in a certain environment, live with a certain group of people, or expose ourselves to certain media, we tend to develop certain behaviors or habits. Our parents and our society heavily influence how we think, feel, and behave. Our habit energies result from the way we have learned to respond to sensory perceptions. These habit energies leave indelible imprints in our mind, and they form internal knots that reside deep in our consciousness. These knots are the blocks of sadness and pain that are tied up deeply in our consciousness. When your mother keeps reminding you that you are fat, that you should not eat this or that, you gradually build up resentment and guilt, forming complex knots of suffering.

When negative habit energies rise up in our mind, they tend to dominate us and limit our horizon so that we cannot see things clearly as they are. Habit energies such as smoking, drinking a lot, and overeating bring us suffering, while habit energies such as humor and generosity bring us joy.

We can easily get attached to our desires—like an insatiable craving for chips—and use them to cope with our emotional hunger. If we are not able to satisfy our craving, knots of desire may form in our mind. Not only the desire for food but also the desire for alcohol, cigarettes, drugs, sex, and praise can lead to the formation of knots in our mind. Once we have experienced one of these pleasant feelings—from getting tipsy, high, and so on—a knot is tied, and we are tempted to seek the same experience again and again. Because of our attachment, we often end up with unpleasant feelings when our cravings are not met, and this causes other knots of suffering to form in our consciousness.

When internal knots of desire, anger, fear, sorrow, and low self-esteem have been buried and repressed in our mind for months, years, or decades, they greatly affect our mental and physical health. The suppression may be the result of wrong perceptions or pressure from societal norms. Because it is easier to avoid suffering in the short term, we have defense mechanisms that push our psychological pains, sorrows, and internal conflicts into our subconscious mind and bury them there. But occasionally they emerge and surface in our thoughts, speech, and actions, reflecting symptoms of physical and psychological disturbance.

If we allow knots to form and then let them grow strong, they will eventually overwhelm us, and it will be very difficult to untie them. It is important that we practice mindful observation and are aware as soon as knots form in us so we can find a way to transform them before they become big and very tightly bound.

The way to transform these afflictions is to look deeply into them. In order to observe them, we need to bring them into awareness through the practice of conscious breathing so that we can recognize our feelings, thoughts, words, and actions as they arise from the depths of our mind.

When we enter a mindful state through conscious breathing, we are essentially closing the doors of our senses. During this time, deeply buried internal knots will have the opportunity to emerge and reveal themselves in the form of images or feelings in our mind. When they surface, we may not be able to understand or see the cause of these unpleasant emotions. But if we shine the light of mindfulness on them, we may see them more clearly. Sometimes the feelings are too intense and unpleasant, and we want to bury them again. But when we are able to maintain and increase the energy of mindfulness, we can overcome our aversion to our painful emotions. We continue to nourish mindfulness through conscious breathing and try to acknowledge our internal knots and conflicts as they emerge. We learn to receive them with the love and tenderness of a mother embracing her baby. We can say: "The light of mindfulness is here and shining, and I know that I have enough strength to be in touch with the knots that are emerging."

It may also be that we need the help of others trained in the practice of mindfulness and capable of dwelling in the present moment to support us in staying with our painful feelings. We can ask a friend to sit next to us and breathe deeply with us as we embrace the feeling together. The collective energy of mindfulness and compassion at a retreat or practice center is also very powerful. Many people are able to transform very deep suffering and release very tightly bound and long-standing internal knots with the help of a loving, supportive spiritual community.

Some years ago Thay (*thay* means "teacher" in Vietnamese, and students often address Thich Nhat Hanh as "Thay") offered a retreat for Vietnam veterans. Many of them carried secrets and tremendous suffering that they had never been able to share with others or find relief from. We would sit in a circle, just listening, and allow each veteran to speak out his suffering. With some of them, we just sat silently for a long time before they were able to open their heart and share with us. One veteran told us that during the war one day his unit attacked guerrilla fighters. They defeated them and brought back the wounded Vietcong to their camp as war prisoners. In his helicopter he transported a woman fighter who was seriously wounded. She was clutching her hammock. Guerrilla fighters living in the jungle slept in hammocks and carried their few possessions with them. She continued to stare at him with great hatred and anger. In her anguished look, he felt she was accusing him: "Why have you come here to destroy my country?" Before they could reach their base, she died in the helicopter, her eyes still staring at him, cold and hard. He had kept her hammock with him all those years and brought it with him to the retreat.

In the retreat we offered teachings on embracing our suffering and holding tenderly our painful emotions. We all trained in mindful walking and breathing, developing our concentration and calm. The veteran began to see that although he had committed terrible acts during the war, he could do very positive and healing things now to heal the wounds he caused in the past. On the last day of retreat we organized a bonfire to help the veterans release their suffering from the war. We practiced walking meditation to the bonfire and encouraged each veteran to throw into the fire objects or symbols that represented his or her pain. The man stood a long time by the fire, clutching his hammock tightly to his chest. He refused to throw it into the fire.

One of the nuns told the man, "Throw the hammock into the fire." But he resisted. He was attached to his suffering, to his complex. Thay went to him and very gently encouraged him to release it. Thay told him, "Now you have become a new person and compassion is born in you. Don't continue to hold on to your old suffering, to your guilt. Give me the hammock." And finally he gave it to Thay. Then the nun and Thay put the hammock into the flames. And there was a big transformation in our friend. He felt so much better: he was light and free from the weight, the complex of guilt that he had carried with him and been attached to for so many years.

By just observing and acknowledging our feelings and our thoughts without judgment, blame, or criticism, we have embarked on the path of emancipation from our suffering. If there is pain, sorrow, or anger, we simply acknowledge that we feel the pain, the sorrow, and the anger. When we acknowledge these feelings with mindfulness, we do not let the feelings of pain, sorrow, or anger take us over and lead us astray.

Instead we try to calm them down with tenderness. Practicing like this will cause our knots to loosen up, and repeated practice will eventually help us understand their roots by identifying the sources of nutriments that have brought them into being. With this insight and understanding, we can stop the suffering at its roots.

The practice is not a matter of transforming our intellect alone. Day and night we have to water the seed of understanding in our store consciousness so that it will grow and help us see the nature of interbeing in everything we see and touch, allowing us to make peace with ourselves. We have to bring this understanding into our daily life so that, with mindfulness, we are more aware of our feelings, recognize them, and prevent them from becoming knots in the first place.

We often confront feelings of regret and fear. We have all had regrets and wished that we had not done a certain thing in the past. If we keep looking back remorsefully, we may create guilt complexes, which prevent us from being happy. We may believe that mistakes have been made already and that we cannot go back to the past to change things. When we look deeply into the relative nature of time, we see that the past has created the present. If we seize the present moment with mindfulness, we are in touch with the past. We can actually go back to the past, while staying firmly rooted in the present moment, and heal the past. We forgive ourselves for our mistakes, knowing we didn't have enough wisdom or the right conditions at that time to do better. We transform our regrets in the present into compassion and understanding, and in this way we also transform the past.

Moreover, since we are the continuation of our ancestors, and we are inextricably linked to them as well as to our parents and our siblings, if we can transform ourselves, we can also help transform them and bring peace and joy to those we love. By taking hold of the present, we can free ourselves from suffering and heal the trauma or wrongdoings that happened in the past.

For example, a mother who has been using fast foods, sodas, and sweets to reward her children's good behavior may feel guilty about her children when they become obese young adults. If the mother lives mindfully, she will see that there are still many opportunities available to help her children master their weight. Furthermore, she can help many other children avoid falling into these unhealthy behavior traps. She can volunteer in schools and participate in their wellness councils, working alongside principals and teachers to ensure that children get healthier food choices at school. In this way, the mother can ease her remorse and feel good that she is contributing to the health and well-being of many other children.

Besides transforming our negative seeds and unpleasant feelings, let us not forget or undermine the strength and wonders of our positive seeds. We have wholesome and unwholesome seeds. However, we need to recognize that each seed is dependent on all the others for its existence. This interdependent nature means that an unwholesome seed contains elements of wholesome seeds and vice versa. We can transform the unwholesome seed simply by watering the wholesome ones. So when we come across difficult times in our lives, if we touch and water the seeds of peace and joy that are already in us, they will sprout and bring the fruits of peace and happiness. Their strong presence will overcome and weaken the unwholesome seeds. Therefore, it is equally important that we regularly observe objects of mind that can lead us to well-being and shine the light of mindfulness on wholesome seeds so that they can grow stronger in the field of our mind.

Practicing Mindfulness in the Twenty-First Century

By practicing these four foundations of mindfulness, we will be able to nourish and protect ourselves, free ourselves from pain, and gain wisdom. Through these practices we penetrate the interrelationship of our physical, physiological, and psychological experience. Furthermore, we can see that each of us and the rest of the world are interdependent—that we as observers, and the world that we perceive, are not separate in time and space. Seeing the true, interconnected nature of our body, feelings, mind, and objects of mind lays the foundation for well-being and happiness.

Generating the energy of mindfulness is essential for transformation. We can live each moment of our life mindfully. We look, listen, think, speak, and act with our mindfulness. When we cook, we cook mindfully. When we eat, we eat mindfully. When we exercise, we move our bodies mindfully. Being aware of our breathing can instantly connect us to what we are doing. By enjoying our breathing in whatever we are doing, we produce the energy of mindfulness to help us touch and live life deeply. This practice allows us to transform the garbage of our afflictions into flowers of well-being.

Every one of us can become a Buddha. A Buddha is someone who is fully awake. Prince Siddhartha Gautama was a human being who became a full-time Buddha after years of concentrated practice. We can all touch Buddhahood as part-time Buddhas when we embark on the path of mindfulness training and the road to attaining healthy weight. Bear in mind that as we practice, we should not view the teachings of Buddhism as rigid dogmas demanding blind commitment. They are simply instruments for gaining insight, which help us remove the obstacles to correct perception.

Being mindful does not mean that we just sit for hours on our meditation cushion in a retreat or monastery. There are many ways to practice mindfulness that can be fully integrated into our daily living. Besides conscious breathing, we can do walking meditation, sitting meditation, smiling, mindful listening, mindful speaking, and mindful working. We can practice concentration and looking deeply in all the activities of our daily life. Even while walking, we can practice stopping. We can walk in such a way that we arrive with each step—not walking just to get somewhere else. We can walk to enjoy each step. If we practice stopping while attending to e-mails, surfing the Web, attending meetings or appointments, folding the laundry, washing the dishes, or taking a shower, we are living deeply. If we do not practice this way, the days and months will fly by without our awareness, and we will lose many precious moments of our life. Stopping helps us live fully in the present. We have many daily opportunities to help our seeds of joy and happiness flourish.

IN CHAPTERS 5, 6, AND 7, we will share more everyday mindfulness practices that you can use throughout your daily life and integrate into almost every task you undertake. Every act and every moment of our life is a valuable opportunity to practice mindfulness.

As we deepen our practice of mindfulness, one breath at a time, one step at a time, we will discover its many wonders. It helps us make real contact with life, making it more meaningful. When we are present, life is also present. Mindfulness practice improves our ability to concentrate. When we can concentrate, we understand and look more deeply into whatever arises. Deep looking ultimately leads to insight and understanding, helping to liberate us from our fear, despair, and suffering and to touch real joy and peace. With mindfulness, we can touch and embrace our life more deeply. We can savor the many gifts that life offers us every day, allowing us to receive nourishment and healing for ourselves and our beloved ones.

PART 2

Mindful Action Plans

CHAPTER 5

Mindful Eating

WE HAVE JUST LEARNED that conscious breathing is an essential practice for bringing our body and mind together, nurturing the well-being of the body and mind, and fostering our connection to all things. And just as mindfully breathing air sustains our physical and spiritual life, so, too, does eating food. Not only does food provide the nutrients and energy we need to support our physical bodies; mindful eating can also help us touch the interdependent nature of all things—and can help us end our difficulty with weight.

Looking deeply at the food we eat, we see that it contains the earth, the air, the rain, the sunshine, and the hard work of farmers and all those who process, transport, and sell us the food. When we eat with full awareness, we become increasingly mindful of all the elements and effort needed to make our meals a reality, and this in turn fosters our appreciation of the constant support we get from others and from nature.

Whenever we eat or drink, we can engage all our senses in the eating and drinking experience. Eating and drinking like this, we not only feed our bodies and safeguard our physical health but also nurture our feelings, our mind, and our consciousness. And we can do this numerous times throughout the day. Mindful eating begins with our choice of what to eat and drink. We want to choose foods and drinks that are good for our health and good for the planet, in the modest portions that will help us control our weight. Yet there are so many types of foods and drinks, and so much information about nutrition and so many diet plans, that we can find it quite difficult to make the right choices. A good way to overcome this challenge is to get up to speed on the latest science-based advice.

The Basics of Eating Well: What Nutrients Are in Our Food?

Food gives the body the raw materials it needs to run the metabolic processes of life. All foods contain at least one, and often two or three, of the so-called macronutrients—carbohydrates, proteins, and fats. These macronutrients give us energy to fuel our daily activities. They also perform unique roles throughout the body. Carbohydrates provide the fastest form of energy, usable by every cell. Proteins provide the building blocks for all of our tissues and organs—skin and muscle, bone and blood, liver and heart. They also form countless cellular mechanisms and minute messengers, such as the enzymes that digest our food and the neurotransmitters that send signals from the brain throughout the body. Fats get woven into the membrane of every cell, insulate nerves, and serve as a precursor to life-sustaining hormones. Food also provides us with vitamins and minerals, the so-called micronutrients—literally, nutrients that are essential in tiny amounts—used to build tissues and catalyze chemical reactions throughout the body.

Nutrition science of the early twentieth century focused on understanding what macronutrients and micronutrients we need to prevent diseases of deficiency, such as protein-deficiency kwashiorkor and vitamin D–deficiency rickets. Starting in the mid-twentieth century, nutrition science shifted its focus to complex chronic diseases such as diabetes, heart disease, and cancer—illnesses that develop imperceptibly over time, have no easy cure, and end lives prematurely. Thanks to many advances in science, we now know quite a bit about what to eat and drink—and what *not* to eat and drink—to prevent these chronic diseases. But you do not need to be a scientist to eat well. Nutrients, after all, are found in foods. And you need only follow a few fundamental food guidelines to maintain your health and well-being.

The dietary recommendations that follow are for adults and are adapted from dietary guidelines developed by experts in the Department of Nutrition at the Harvard School of Public Health.[1]

Carbohydrates, Proteins, and Fats: Choosing the Healthiest

Diet books from Atkins to Zone portray carbohydrates as the enemy. Other nutrition gurus tout low-fat, high-carbohydrate diets to lose weight and prevent disease, or push high-protein diets as the way to achieve good health and a healthy weight. The truth about macronutrients and health, however, is that the *type* of carbohydrate, protein, and fat we choose is much more important than their relative amounts in our diet.

Take carbohydrates. They are found in many types of whole and processed foods—from apples to ziti—but not all carbohydrates are created equal. The healthiest carbohydrates come from whole grains, legumes, vegetables, and whole fruits. The least-healthy carbohydrates come from white bread, white rice, pasta and other refined grains, sugary foods and drinks, and potatoes. We talk about the reasons for limiting these less-healthy carbohydrates later in this chapter. Here, let's focus on the positive: Whole grains, vegetables, whole fruits, and legumes are good choices as carbohydrates, and they are also rich in vitamins, minerals, and fiber. Whole-grain foods—such as whole-wheat bread, whole oats, brown rice, millet, barley, quinoa, and the like—deserve special mention, because more and more research points to the benefits of making whole grains a daily habit. Long-term studies have found that people who eat, on average, two or three servings of whole-grain foods a day have a 20 to 30 percent lower risk of heart disease and diabetes, compared with people who rarely eat whole grains.[2] Eating whole grains may also offer some protection against colon cancer, but more research is needed on this diet-to-disease relationship.[3]

Why Are Whole Grains So Good for Your Health?

Exactly how whole grains protect against heart disease and diabetes is still an open research question. What we do know is that whole grains contain fiber, which slows their digestion and makes for a gentler rise in blood glucose after a meal; the soluble fiber in whole grains, especially the type found in oats, also helps lower low-density lipoprotein (LDL), the "bad" cholesterol. The germ in whole grains provides folate and vitamin E, and whole grains are also a source of magnesium and selenium—vitamins and minerals that may help protect against diabetes, heart disease, and some cancers. Yet some studies have found that the benefits of whole grains go beyond what can be attributed to any individual nutrients that they contain. What seems more likely is that whole grains' health benefits accrue from their special combination of nutrients.[4] The whole is truly greater than the sum of its parts—an aspect of interdependent nature.

It's a similar story with proteins. Plant foods or animal foods can all provide the body with the protein it needs. But when choosing foods high in protein, we must pay attention to the other nutrients that travel along with the protein. The healthiest plant sources of protein—beans, nuts, seeds, whole grains, and foods derived from them—also contain fiber, vitamins, minerals, and healthy fats, and they are earth-friendly selections, too. Among the animal sources of protein, some contain healthful fats (fish) or are relatively low in harmful fats (chicken, eggs). (For more information on choosing healthy fats, see table 5.1 on page 105.) But red meat and full-fat dairy products are high in a type of fat that is bad for our hearts; furthermore, as we discuss later in this chapter, consuming red meat, processed meat, and dairy products may increase the risk of some cancers. Red meat and dairy products also take a terrible toll on the environment. So to pick the healthiest sources of protein, both for your own welfare and for that of the planet, choose plant-based proteins from nuts, legumes, seeds, and beans. If you do have to consume animal foods, choose fish or chicken. If you have to eat red meat, it's best to limit yourself to no more than once or twice a week. Whole eggs can be a healthful source of protein but should be consumed in moderation, since eating an egg or more a day may increase the risk of diabetes and may increase the risk of heart disease in people who have diabetes;[5] if you have heart disease or diabetes, you should eat less than that amount per week.

Vegetarians and Protein: Variety Is Essential to Good Health

There is one difference between plant and animal proteins that is important for us, and especially for vegetarians, to understand. Our body takes the protein in plant and animal foods and breaks it down into smaller components, called amino acids, which it then uses to build and restore tissues and to run a multitude of functions. Some amino acids are "essential," meaning that the body cannot make them and must obtain them from food. Others are not essential, and the body can build them by rearranging the essential amino acids. Proteins from animal foods are called "complete proteins," meaning that they contain all the essential amino acids. Proteins from plant foods are called "incomplete proteins," meaning that they tend to be low in one or more essential amino acids. Even so, plant proteins can meet your daily protein needs, as long as you choose a variety of plant foods and get enough calories throughout the day. So vegetarians should take care to eat varied high-protein vegetarian foods every day—beans (including tofu), nuts, seeds, and whole grains—to ensure that they get enough of all the essential amino acids.[6]

The same "quality matters more than quantity" message is true about fats. Some fats are so beneficial that you can enjoy them every day, while others are so harmful that you should severely limit them or avoid them altogether. There is an easy way to tell healthy fats from unhealthy fats. Most of the healthy fats—the monounsaturated and polyunsaturated fats—come from plants and are liquid at room temperature. Rich green olive oil, golden sunflower oil, the oil that rises to the top of a jar of natural nut butter, and the oils that come from fatty fish are all examples of healthy unsaturated fats. The unhealthy fats—saturated fats—and the *very* unhealthy fats—trans fats—tend to be solid at room temperature, such as the fat that marbles a steak or that is found in a stick of butter or margarine. Meat and full-fat dairy products are the biggest sources of saturated fat in the Western diet; tropical palm and coconut oils are also high in saturated fat. The trans fat in the Western diet comes primarily from vegetable oils that have been partially hydrogenated, a chemical process that makes oils more solid and stable at room temperature—and makes them extremely harmful to our health.

What effect do these different types of fats have on our health? Numerous studies have found that when people replace carbohydrates in their diet with monounsaturated and polyunsaturated fats, their blood-cholesterol profile improves—heart-harmful LDL cholesterol goes down, and protective high-density lipoprotein (HDL) cholesterol goes up.[7] Saturated fats, meanwhile, cause both HDL and LDL to rise, so unsaturated fats are a better choice for heart health. Trans fats are the worst kind of fat, harmful in even small amounts.[8] They drive down protective HDL and cause a rise in damaging LDL, and they damage the cells that line our arteries. Research also suggests that trans fats trigger inflammation,[9] a red alert in our immune system that may underlie a number of deadly diseases, including heart disease, stroke, and possibly diabetes. Diets high in trans fats may promote weight gain,[10] although more research is needed into the relationship between trans fats and obesity. Furthermore, eating a low-trans-fat diet that is high in healthy fats may lower the risk of age-related macular degeneration.[11]

Trans fats are getting a bit easier to avoid: since word has gotten out about their ill effects—and since manufacturers have been required to list them on their food labels in the United States—many food manufacturers and restaurants have begun eliminating them from their products. It is nearly impossible to avoid all saturated fats, however, since even healthful sources of unsaturated fats—such as peanuts and olive oil—contain a small amount of saturated fat. So for good health, enjoy healthy fats, limit saturated fat, and avoid trans fat. (See table 5.1.)

To Control Your Weight, Calories Matter

While there is increasing evidence about the best carbohydrate, protein, and fat choices for health, there has been ongoing debate about the best choices for weight loss. Of course, to lose weight, dieters need to eat fewer calories than they burn. The big question has been whether the relative amounts of carbohydrate, protein, and fat in the diet hold any particular advantage for calorie control and weight loss. Some scientists advocate a low-fat diet, while others stand by a lower-carbohydrate approach, or a Mediterranean-style eating plan, with moderate amounts of healthy fats and plenty of fruits, vegetables, and fiber. Two well-designed clinical studies with hundreds of participants that put these competing diet styles to the test have come up with similar conclusions: people can lose weight using any of these different strategies as long as they lower the amount of calories that they consume; and having social support for making these behavior changes may help them succeed.[16]

So to achieve a healthier weight, the message is to find a lower-calorie eating plan that you can follow—one that lets you consume healthy foods that you enjoy—and to find some support for following it. Some people may find that support in a formal weight-loss program; some may find it from an online community; some may choose to create that support within their families or circle of friends, joining or starting a mindfulness living *sangha,* or by working with colleagues to get healthier foods added to the company cafeteria. (To find a mindful living *sangha* in your area, visit www.iamhome.org.)

You may be wondering how many calories you should consume each day to maintain your weight and how much to cut back to lose weight. There's no one answer to that question, since calorie needs vary depending on age, gender, body size, and level of physical activity. Some people may need only 2,000 to 2,500 calories per day to maintain their weight, and a bit less to lose weight, while others who have a larger body size or are very active may be able to eat more calories yet still lose weight. As people lose weight, their daily calorie needs drop. There are many Web sites that offer calorie-need calculators based on your current weight and your weight goals, and consulting a health professional, especially a registered dietitian, about your calorie needs is also useful. As a practical guide, an energy deficit of about 250 to 500 calories per day can result in two to four pounds of weight loss per month, and a safe approach to creating this energy imbalance is to moderately reduce one's calorie intake and increase one's activity. This amounts to cutting back on sugary soda by about a can a day and adding a brisk walk daily. (See chapter 6 for more advice on physical activity and weight loss.)

Table 5.1 GOOD FATS, BAD FATS, AND VERY BAD FATS

What type of fat is it?	What foods contain it?	How should you consume it?
Monounsaturated	Olive oil, canola oil, peanut oil, sesame oil, peanuts, almonds, sesame seeds, avocado, nut butters	Enjoy
Polyunsaturated	Safflower oil, sunflower oil, corn oil, soybean oil, sunflower seeds, walnuts, flaxseeds, flaxseed oil, fish, tofu	Enjoy
Saturated	Animal products, especially red meat; full-fat dairy products such as milk, cheese, butter, and ice cream; tropical oils (palm, coconut*)	Limit
Trans	Partially hydrogenated vegetable oils (found in some** hard-stick margarines and some soft margarines, commercially prepared baked goods, deep-fried foods, fast food, and restaurant food)	Avoid

*Coconut oil is less harmful than other forms of saturated fat, since it raises HDL ("good" cholesterol), so it is fine to include a small amount of coconut oil in your diet.
**Margarine can be a healthful fat choice as long as it contains no trans fats and no partially hydrogenated oils, so check the nutrition-facts label for zero trans fats, and check the ingredients list to make sure the margarine contains no partially hydrogenated oil.

Omega-3s—An Especially Healthy Fat

When you choose your healthy fats, make a special effort to include omega-3 fats, polyunsaturated fats that are extremely beneficial for the heart. Omega-3 fats appear to protect against erratic heart rhythms that may cause sudden death.[12] They may also hold benefits for people suffering from inflammatory diseases such as rheumatoid arthritis.[13] Omega-3 fats are essential fats, meaning that our bodies cannot manufacture them and we must obtain them from food or supplements. So it's a good idea to enjoy at least one omega-3-rich food every day.

An earth-friendly source of omega-3s are the plant foods that contain alpha linolenic acid (ALA)—principally, walnuts, canola and soybean oils, flaxseeds and flaxseed oil, dark leafy greens, and chia seeds (also known as salvia). For those who consume fish, fatty fish—such as salmon, tuna, bluefish, and mackerel—are rich in two types of omega-3 fats, eicosapentaenoic acid (EPA) and docosahexaenoic acid (DHA), also known as long-chain omega-3 fats. Much of the research on omega-3s and heart disease has focused on consumption of fish or fish-oil supplements.[14] Our body can convert the plant form of omega-3 into the long-chain form, but it does so slowly, and there has been much scientific debate over whether these different forms of omega-3 fats work the same beneficial way in the body. Vegetarians, however, can take heart: newer research shows that plant omega-3s may also play an important role in protecting the heart, especially in people who do not regularly eat fish.[15]

Eat and Drink for Your Health and Our World: Practical Guides

Beyond the technical words like *carbohydrates, proteins,* and *fats,* there are a few practical guidelines to follow that will steer your eating habits in a healthier direction to control weight and help you lower your risk of diseases.

Go with Plants

A mindful diet for weight loss should first and foremost be a healthy diet—both for you and for the planet. And the first and most essential principle of healthy eating is to shift more to a plant-based diet. Asians in particular have practiced vegetarianism for thousands of years. As we discussed in chapter 3, the ethical and environmental arguments for a plant-based diet are stronger than ever. The health benefits for eating a plant-based diet are equally strong.

Decades of research on hundreds of thousands of men and women has shown that eating a diet rich in vegetables, fruits, whole grains, and healthy fats and low in refined grains and unhealthy fats can lower the risk of heart disease and diabetes.[17] Some research even suggests that people who eat little or no meat may live longer than people who follow a more meat-heavy diet[18] (although the science isn't definitive and not all studies have found such a mortality benefit[19]). Vegetarians and vegans tend to weigh less and have lower blood pressure, lower blood cholesterol, and, in turn, a lowered risk of heart disease than people whose diets include some or all types of animal products;[20] they may also have lower risks of some cancers, though the studies are conflicting and more research is needed.[21] (Of course, for optimum health, vegans must take care to get adequate vitamin B_{12}, vitamin D, and other nutrients they may be missing by avoiding animal foods.)[22]

There's also strong evidence of the health hazards associated with eating animal foods. The Center for Science in the Public Interest estimates that the saturated fat and cholesterol in red meat, poultry, dairy products, and eggs cause sixty-three thousand heart-disease-related deaths a year in the United States and another eleven hundred deaths a year from food poisoning.[23] People who eat meat and processed meat have a higher risk of diabetes than individuals who follow a vegetarian diet.[24] High levels of red-meat consumption, and any level of processed-meat consumption, raises the risk of colon cancer,[25] and eating meat, especially meat that is cooked to a high temperature, may increase the risk of pancreatic cancer.[26] The Nurse's Health Study II, meanwhile, followed nearly forty thousand women for seven years to determine the relationship between red-meat consumption and the risk of getting early breast cancer. It found that for every additional 3.5 ounces of red meat consumed per day—a portion of meat about the size of a medium fast-food hamburger—the risk of premenopausal breast cancer rose by 20 percent.[27]

You do not need to become a 100 percent vegetarian to achieve the health benefits of a plant-based diet. Several studies have shown that following a "prudent" diet pattern—one that is rich in vegetables, fruits, whole grains, and healthy fats but does include fish and poultry—rather than a meat-heavy diet may lower the risk of several deadly and disabling diseases, among them diabetes,[28] heart disease,[29] stroke,[30] and obstructive lung disease,[31] as well as lower the risk of dying from heart disease, cancer, or any other cause.[32] A similar line of research has found evidence that following a Mediterranean-style diet pattern, which is also plant based but includes dairy and fish, can lower the risk of heart disease, stroke, Parkinson's and Alzheimer's disease, and cancer, as well as the risk of dying from heart disease, cancer, or any other cause.[33] So you can benefit from becoming even a part-time vegetarian.

Fill Your Plate with Vegetables of All Different Colors—And Enjoy Your Fruits Whole

When it comes to vegetables and fruits, the basic message comes down to two words: eat more. People who eat diets rich in vegetables and whole fruits may lower their blood pressure as well as their risk of heart disease, stroke, diabetes, and possibly some cancers.[34] Diets rich in vegetables and fruits may lower your risk of cataracts and macular degeneration,[35] and thus help protect your vision as you age.

The benefits of eating whole fruits and vegetables likely accrue from the nutrients that they provide, as well as from the absence of less-healthy or higher-calorie foods, which they may replace on your plate. Fruits and vegetables are loaded with vitamins, such as vitamin C, which gives a boost to the immune system and also acts as a powerful antioxidant, preventing cellular damage from free radicals; vitamin K for strong bones; and beta-carotene, a precursor to vitamin A and also an antioxidant. They are rich in minerals, including potassium, which may help lower blood pressure, and magnesium, which may help control blood glucose. They are also a great source of healthy carbohydrates, including fiber. Special plant chemicals, also known as phytochemicals, that give vegetables and fruits their bright colors may also play beneficial roles in protecting against disease. Lycopene, for example, a pigment that helps make tomatoes and watermelon such a vibrant red, may protect against prostate cancer. Lutein and zeaxanthin, other members of the carotenoid family, may help prevent age-related macular degeneration.

To get the benefit of all these protective nutrients, make an effort to choose vegetables and fruits in a rainbow of colors every day. Include dark green varieties, such as broccoli, kale, Brussels sprouts, and collard greens; yellow-orange, such as sweet potato and apricots, carrots and cantaloupe; red, such as tomatoes, watermelon, strawberries, and red bell peppers; white, such as onions, garlic, and cauliflower; and purple-blue, such as red cabbage, beets, and blueberries. Make it your goal to consume at least five servings of vegetables and fruits a day, since several studies find that the heart-healthy benefits of vegetables and fruits begin to accrue at this level.[36] More is certainly better. A serving is about half a cup of cooked vegetables or chopped fruit, or one cup of salad greens. To make it easier to gauge the portion, devote half of your plate to vegetables or fruits at each meal.

Remember to enjoy your fruit whole, not drunk as a big glass of juice. Fruit juice—even 100 percent fruit juice—is high in rapidly digested sugar. A glass of orange juice has as much sugar and calories as a glass of Coca-Cola. Fruit juice also lacks the fiber of whole fruit. Indeed, the Nurses' Health Study found that women who drank a cup or more of fruit juice per day had a 40 to 50 percent higher risk of diabetes than women who drank fruit juice less than once a month. Eating whole fruit, however, was associated with a lower risk of diabetes.[37]

To extend the benefits of fruits and vegetables beyond your own health, buy your fruits and vegetables from a local farmers' market, or buy a share in a community-supported farm. You'll be supporting your local economy, you'll enjoy fruits and vegetables at the peak of their freshness, and your produce will consume fewer fossil fuels on its way from the farm to your plate. (See the appendix for information about how to find community-supported farms.)

Limit Potatoes, Refined Grains, and Sweets

You may notice one vegetable that is conspicuously absent from the list of rainbow-colored vegetables: the potato. While multiple studies have shown the benefits of eating fruits and vegetables, potatoes do not seem to play a role in these observed protective effects. That's because potatoes—whether their skins are brown, red, yellow, or purple—have more in common with white bread and white rice than they do with broccoli or bell peppers. Potatoes contain rapidly digested starch, and large amounts of it.

Eating a large portion of such starchy foods can send your blood sugar on a roller coaster. First, as your body quickly converts the starch to glucose and absorbs the glucose from the gut, blood-sugar levels rise high; your pancreas pumps out insulin to rapidly clear the glucose from the blood, but it may overshoot things a bit, causing your blood sugar to dip a bit lower. This sequence of events may lead you to feel hungry again, not long after finishing your meal. Over time, eating diets high in such rapidly digested starchy foods may increase your risk of heart disease and diabetes,[38] and there is evidence that limiting these types of foods in your diet may help with weight loss.[39] So eat potatoes sparingly, if at all, and when you do, don't count them as part of your five-plus servings of vegetables a day.

Eating large amounts of refined grains and sweets, similar to eating lots of potatoes, can cause a rapid rise in blood sugar, a spike in insulin, and then an equally precipitous dip in blood glucose. Furthermore, the refined grains that fill our supermarket shelves—white rice, white bread, white pasta, and anything made with white flour—are nutritionally bankrupt substitutes for whole grains. The grain-refining process removes the bran and the germ, taking away nearly all of the fiber and many of the beneficial vitamins and minerals. What's left is the starchy middle, or endosperm. Food manufacturers must, by law, add back some of the lost nutrients to refined grains, but they do not replace everything that has been stripped away.

The American Heart Association (AHA) has recommended that Americans drastically cut back on added sugar, to help slow the obesity and heart disease epidemics.[40] The AHA's suggested added sugar threshold is no more than 100 calories per day (about 6 teaspoons or 24 grams of sugar) for most women and no more than 150 calories per day (about 9 teaspoons or 36 grams of sugar) for most men. Keep in mind, however, that your body doesn't need to get any carbohydrate from added sugar. A good rule of thumb is to skip products that have added sugar at or near the top of the list, or have several sources of added sugar sprinkled throughout the list.

Take a Daily Multivitamin with Extra D to Give You a Nutritional Safety Net

If you live in the higher latitudes, spend a lot of time indoors, have a dark skin tone, or are overweight or obese, you may be deficient in vitamin D without even realizing it. Vegans and others who mindfully limit their intake of animal products may also be deficient in some nutrients, such as vitamin B_{12}. That's why nutrition experts at the Harvard School of Public Health recommend that adults take a daily multivitamin as "nutrition insurance."

There's no need to buy a fancy supplement. Even a standard store-brand supplement will have enough of the basic vitamins and minerals that you need. There's also no need to take a supplement that provides more than 100 percent of the daily value of any vitamin or mineral, with the exception of vitamin D, a nutrient that is critical for bone health and that scientists believe may also play a role in preventing chronic diseases such as heart disease, some cancers, infectious diseases, and multiple sclerosis.[41] One billion people worldwide are thought to be deficient in vitamin D, and scientists now think that our daily vitamin D needs are much higher than once thought.[42] Few foods are naturally rich in vitamin D, and even foods that are fortified with vitamin D (such as milk in the United States) do not provide much of it. Furthermore, during the winter months, the bodies of people who live in higher latitudes can't make enough vitamin D from exposure to the sun. That's why many people may benefit from taking 1,000 to 2,000 international units (IUs) of supplemental vitamin D a day. Since a standard multivitamin typically provides only 400 IUs, you may want to ask your doctor to evaluate whether you need a vitamin D supplement in addition to your multivitamin.

Finally, be sure to look for a multivitamin that derives most if not all of its vitamin A from beta-carotene rather than from retinol. Consuming high levels of retinol may increase the risk of fractures; pregnant women should also avoid taking high levels of retinol, as this may lead to birth defects.[43]

Limit Sodium

Sodium is an essential nutrient, but most of us get far more of it each day than we need. High-sodium diets can exacerbate high blood pressure in some individuals. Reducing sodium can lower blood pressure and, over the long term, can also lower the risk of heart attack and other cardiac problems.[44] It's best to limit sodium to less than 2,300 milligrams per day—the amount found in about one teaspoon of table salt. People who have high blood pressure or are at risk of high blood pressure (including people over the age of forty, African Americans, or people who have prehypertension) should cut back further, to no more than 1,500 milligrams per day. Indeed, the AHA now recommends that most adults cut back to 1,500 milligrams of sodium per day, since new research estimates that 70 percent of U.S. adults fall into this high-risk, salt-sensitive group.[45]

One way to reduce the sodium in your diet is to cut back on processed foods. Food manufacturers add loads of sodium to frozen meals, soups, condiments, cheese, bread, and chips to cater to our taste for saltiness, but also to improve texture and extend shelf life. Fast-food and sit-down restaurants also offer exceptionally salty fare; the Center for Science in the Public Interest, for example, has found that some U.S. restaurant appetizers and entrées contain more than a day's worth of sodium.[46]

Cutting back on processed and restaurant foods may also help you limit the amount of another high-sodium food additive, the flavor-enhancing monosodium glutamate (MSG)—and emerging research suggests MSG consumption may be related to weight. A small study in China found that people who had the highest MSG intake were nearly three times as likely to be overweight as people who had the lowest MSG intake.[47] The findings are preliminary, and researchers have yet to tease out the way that MSG may be related to weight. It's possible that the enhanced flavor of MSG-laced food spurs people to simply eat more of it; it's also possible that MSG has an effect on the brain centers or hormones that control hunger.[48]

Get Enough Calcium—But Consider the Source

Calcium—the mineral that is essential for strong bones and teeth, the steady beating of the heart, and countless other bodily functions—has been the focus of much scientific debate. The U.S. government recommends that we consume 1,000 milligrams of calcium per day, while the United Kingdom recommends only 700 milligrams per day. Some critics speculate that the U.S. recommendations are shaped more by the lobbying of the powerful dairy industry than they are by the scientific evidence.[49]

There's also been debate about how best to get calcium. The U.S. Department of Health and Human Services and the Department of Agriculture's Dietary Guidelines for Americans recommends that adults consume three glasses of milk a day. Yet milk and dairy foods are high in unhealthy saturated fat. Even fat-free milk has about eighty calories per glass, and three glasses can really bust the calorie budget of someone trying to lose weight.

This debate over milk gets even more complicated when one considers the relationship between dairy, calcium, and chronic disease. Dairy consumption can protect against colon cancer in modest amounts;[50] high levels of dairy intake on their own, however, do not seem to offer protection against fractures late in life.[51] If there were no harm in consuming high amounts of dairy and calcium, this would be purely an academic debate. But studies raise the disturbing possibility that high levels of milk or calcium intake are associated with increased risk of prostate cancer in men and that high levels of lactose intake are associated with increased risk of ovarian cancer in women.[52]

Milk production also has a huge environmental impact, as we described in chapter 3. And there is an ethical implication of drinking milk, since the treatment of cows on dairy farms is often not very compassionate and cows who can no longer produce milk are then slaughtered. The dairy and beef industries are closely linked, so even if you do not consume the meat of cows or wear leather made from them, someone else does.

So what is the best way to get calcium? If you plan your diet carefully, you can get enough calcium from nondairy sources—among them, leafy green vegetables, calcium-set tofu, and tahini. For some people who eat a vegan diet, however, they may want to consider taking calcium supplements or making sure to consume calcium-fortified soy, grain, or nut-milk beverages, a small glass of calcium-fortified juice, or calcium-fortified cold cereals. If you do want to consume dairy products, having a modest amount—no more than one or two servings per day—and eating a healthy diet rich in vegetables and legumes can also provide adequate calcium. An additional benefit of taking calcium supplements is that they are often fortified with vitamin D, which helps calcium absorption.

Choose Healthy Drinks

Water is the best drink choice for health and weight loss. Sugary drinks are the worst choice, since consuming them in excess contributes to the risk of obesity, diabetes, and possibly even heart disease.[53] Sometimes, however, it is not that obvious that a beverage is high in sugar and calories. If you mindfully read the nutrition-facts label, you will see that natural "100 percent fruit juice" has as many calories and as much sugar as a soda. Grape juice and cranberry juice cocktails have *more* sugar and calories than a soda. If you enjoy juice, stick to one small glass a day, about the size of an old-fashioned "juice glass" (4 to 6 ounces). Energy drinks and sports drinks also contain lots of sugar, though drink marketers often try to disguise these beverages as "healthy" by boasting about the vitamins, electrolytes, antioxidants, or herbs they contain. Don't be fooled. Keep in mind that there are many different types of sugar added to drinks—cane sugar, honey, high-fructose corn syrup, fruit-juice concentrates—but to the body, they are all sources of extra calories and sugar. Diet drinks, sweetened with artificial sweeteners, may not be the best alternative, since it is unclear what their long-term effects are on weight and health.

The Department of Nutrition at the Harvard School of Public Health has developed a "traffic-light" system for ranking beverages (see figure 5.1).[54] Those highest in sugar—sugary sodas, fruit juices, smoothies, and sports drinks—fall into the "red" category: "drink sparingly and in-

Figure 5.1 HOW SWEET IS IT?

Calories and teaspoons of sugar in 12 ounces of each beverage

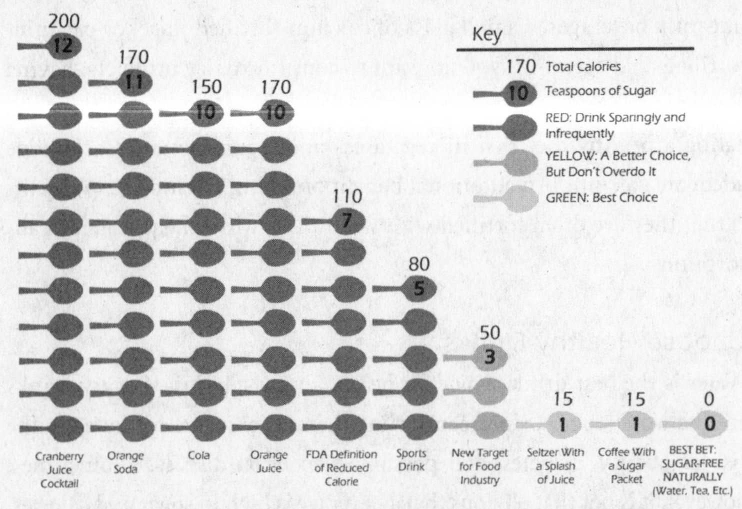

Copyright © 2009 Harvard University. For more information, see The Nutrition Source, Department of Nutrition, Harvard School of Public Health, http://www.thenutritionsource.org.

frequently, if at all." Slightly sweet beverages—those that have no more than one gram of sugar per ounce and are free of artificial sweeteners—fall into the "yellow" category: "a better choice, but don't overdo it." The "green" beverages are your best bet—beverages that are sugar-free naturally, such as water or sparkling water. Tea or coffee can be a healthy choice for most people, in moderation (up to three or four cups a day) and may even have some health benefits.[55] Skip the sugar and cream to keep these beverages low-calorie and healthful. Pregnant women may want to limit their caffeine. People who get jittery or have sleep problems when they consume caffeine may also want to limit caffeine.

Be Mindful of Alcohol Consumption

From a health perspective, alcohol should be limited, if you consume it at all. Heavy drinking increases the risk of chronic diseases such as hypertension; cirrhosis; esophageal, breast and colon cancers; and alcoholism.[56] Alcoholism also takes a tremendous toll on families, communities, and nations. Worldwide, alcohol consumption has been estimated to cause one in twenty-five deaths, and the cost of alcohol to society—an estimated 1 percent of gross domestic product in high- and middle-income countries—includes the cost of both alcohol-fueled chronic diseases and alcohol-related social problems.[57] Although moderate drinking can lower the risk of heart disease and diabetes,[58] it can also raise the risk of breast cancer and colon cancer.[59] If you do not drink, there's no reason to start drinking, since there are many other ways to improve your heart health and lower your risk of diabetes. (Skipping sugary drinks is one way; exercising more is another.)

Scientifically, moderate drinking is defined as no more than one drink per day for women and no more than two drinks per day for men, and until middle age, the risks outweigh the benefits.[60] Yet even drinking in moderation can be the beginning of alcoholism for some people. Although you yourself may always be on guard about not becoming addicted to alcohol and can safely drink a glass of wine occasionally without overindulging, this may not be the case for your children, your grandchildren, and other loved ones. Every time you drink in front of them, you may be increasing the likelihood that they will drink in the future. And they may grow up and become dependent on alcohol. By abstaining from alcohol completely, you become a role model for them and may help to protect them from turning to alcohol as a habit or in times of stress and difficulty. Alcohol is an addictive substance.[61] Having the first glass may lead to the second and the third. When you see that as a society, we are in great danger because of how we use alcohol, refraining from the first glass of wine is a manifestation of your enlightenment. You do it for all of us.

We must look deeply to see that when we practice mindful consumption we practice not only for ourselves but for others. The way you live your life is for your ancestors, future generations, and your whole society. Even if we completely refrain from all drinking, we can still get killed by a drunk driver, so to help one person stop drinking is to make the world safer for everyone. When we can be free from the shell of our small self and see our interrelatedness to everyone and everything, we understand that our every act is linked with all of humanity, with the whole cosmos. To stay healthy is to be kind to your ancestors, your parents, future generations, and also your society.

From the standpoint of mindfulness and compassion, we encourage you to abstain from alcohol. As in the case of meat consumption, reducing alcohol consumption can impact world hunger, as the grains and foods used in alcohol production can be used instead for direct human consumption. If you are unable to completely stop drinking, then at least reduce the amount you drink by one-third, one-half, or two-thirds. No one can practice perfectly, including the Buddha. Even vegetarian dishes are not entirely vegetarian. Boiling vegetables kills the bacteria living in them. Although we cannot be perfect, because of the real danger alcoholism poses to our society, destroying many families and causing much suffering, we should practice to reduce or completely stop our alcohol consumption. We have to live in a way that will avoid the tragedy alcohol abuse can create. This is why even if you can be very healthy while enjoying one glass of wine every week, we still urge you to look deeply into the detrimental effects of alcohol on our society and do the best you can to minimize alcohol consumption. Whenever you drink alcohol, ask yourself: Do I really want to drink this? And if you are going to drink, drink mindfully.

The Practice of Mindful Eating

Now that we've covered the basics of healthy eating, let's focus on how to eat mindfully so that we truly enjoy our food and eat with compassion and understanding. Mindful eating means simply eating or drinking while being aware of each bite or sip. You can practice it at any meal, whether you are alone in your kitchen or with others in a crowded restaurant. You can even practice mindful drinking when you pause to take a sip of water at your desk. Mindful eating allows us to fully appreciate the sensory delight of eating and to be more conscious of the amount and nature of all that we eat and drink. When practiced to its fullest, mindful eating turns a simple meal into a spiritual experience, giving us a deep appreciation of all that went into the meal's creation as well a deep understanding of the relationship between the food on our table, our own health, and our planet's health.

Engaging in mindful eating, even if only for a few minutes, can help you recognize how the practice of mindfulness encompasses all spheres and activities, including ordinary tasks. Take drinking a glass of water: if we are fully aware that we are drinking the water, and we are not thinking of anything else, we are drinking with our whole body and mind. While eating, we can also be aware of how we feel and of how we consume, whether we are truly hungry, and whether we are making the best choices for our health and the health of the planet.

Mindful eating sees each meal as representative of the whole cosmos. Recall the apple meditation from chapter 2. Look closely at an apple and you can see a cloud floating in it, as well as the rain, the earth, and the sunshine that made the apple tree flower and fruit. Recognize that this apple contains the universe. When you bite into the apple, can you be fully aware that this is a miracle from the universe that you have just put into your mouth? Notice that there is nothing else in your mouth as you chew, no worries or anxiety. When you chew the apple, just chew the apple, not your future plans or anger. You must chew very consciously and with focus. When you are able to be there for the apple 100 percent, you will feel connected to the earth, to the farmer who grew the apple, and also to the person who brought it to your table. Eating this way, you feel that strength, freedom, and pleasure are attainable. This meal nourished not only your body but also your mind—your whole being. Let's see how a mindful chef would apply these practices in the kitchen and at the dinner table.

Chef Sati Invites You to Dinner

Chef Sati—a chef and Buddhist teacher—invited us to dinner to introduce us to the art of mindful cooking and eating. He promised that his dishes would enlighten our senses and that the evening would simply be an opportunity to touch the joy of life, but we would all have to take an active part in the meal.

He placed an array of colorful vegetables, whole grains, and spices on the counter. As we washed the vegetables, he said, "In these vegetables, I see the sun, the earth, the clouds, the rain, and numerous other phenomena, including the hard work of the farmers. These fresh vegetables are gifts of the universe. Washing them, we know we are also washing the sun, the earth, the sky, and the farmers." Because we were attentive to what we were doing, we touched the interdependent nature that makes life possible and felt deep joy to be living in the present moment.

The host made a wonderful soup with mushrooms; a salad bursting with red, yellow, orange, burgundy, and green colors; oven-fresh wholegrain bread with nuts; and, along with other dishes, what appeared to be a chicken dish with a wonderful sauce, decorated with green onions and cilantro. The kitchen was filled with brilliant colors and fine textures of various ingredients, aroma from different herbs and spices, sizzling and bubbling music from the stove—all leading to the mouthwatering dishes that were masterfully prepared. All of our senses came alive. Working together, we prepared all the dishes in no time at all.

As we sat down around the dinner table, we looked at the empty plates in front of us that we were eager to fill with the scrumptious food that Chef Sati had assembled in the middle of the table. He gently exclaimed, "It's so wonderful for us to be together, and I am so grateful we can share this dinner because in many other parts of the world, our empty plates might remain empty for a long time. Eating is a very deep practice. Let's learn to eat with compassion and understanding." There was plenty of food on the table, but he suggested that we start by taking only a small portion first. Following his cues, before we served ourselves we smiled at each other, looked at the food, and smiled to it with gratitude. When we picked up our piece of bread, we did not devour it immediately. Behind the tenderness and the inviting aroma of the bread, we tried to look deeply into it—to see the sunshine inside, the cloud inside, the earth inside. We saw also all past generations of wheat plants leading up to the wheat of our times, as well as all the many generations of farmers and scientists who contributed to the evolution of the wheat and bread industry. The bread carried a lot of love and care of many people, including our dinner host. When we had seen clearly the real piece of bread, we put it into our mouth and chewed it mindfully—chewing and tasting only the bread, and not the worries in our minds. This way, we truly enjoyed the bread and could fully receive it as a gift from the universe. The piece of bread in our mouth was a miracle, each of us was a miracle, and the present moment of being together was also a miracle.

As we slowly ate the bread, soup, salad, and other dishes, we became fully aware of each bite, awakening all of our senses. We started to deepen our relationship with the food and to feel how we were all connected in a miraculous way.

Knowing the long history of vegetarianism in Buddhism, we asked our host about the chicken dish. He said, "This may be a surprise to you, but the chicken dish was not prepared with real chicken meat. It is made with soy and wheat gluten, by infusing the glutinous protein with spices and sauces to simulate the flavor and texture of chicken. I included this dish to ensure there would be enough protein in the meal." The dish was indeed as flavorful as it looked. He further explained, "I don't use real chicken meat or any other kinds of meats or animal products in cooking because it harms the environment and is not compassionate toward animals." The wholesomeness of the chicken dish was truly amazing, and we enjoyed every bite of it. We felt fortunate that we could sit down and enjoy a meal like this, being so present to the food and our friends.

After we finished our meal and had all commented on how good it felt to eat in this peaceful, mindful way, our host unabashedly asked us to help wash the dishes. We knew, of course, that there was another valuable mindfulness experience coming. The idea of doing the dishes was not exactly fun. But once he stood in front of the sink and tenderly and joyfully cleaned the plates with as much care as he had cooked all the wonderful dishes, we rolled up our sleeves and took turns washing as well. He explained that we needed to be mindful of everything we did, including the most mundane of chores such as dishwashing. We should be fully aware of the plates, the water, the amount of detergent, and each hand movement with the sponge. We could wash each dish with the same care that we would use if we were bathing a baby. We had to pay full attention to the task. Any distraction that disengaged us from the dishes would take us away from experiencing the present moment, a valuable moment in which we and the dishes are related. Sounding like a light breeze, he offered, "We should do the dishes just to do the dishes. We should live fully in the moment while washing them, not think about dessert or about going home. Otherwise, we miss an important appointment with life."

After we cheerfully finished putting away the dishes, we made a light, fragrant tea to accompany our dessert, which was a composition of fresh green, yellow, and red melons carefully cut in triangular wedges and presented beautifully on each plate. Chef Sati showed us how to drink tea. He held a cup of tea in his hands, breathed in the fragrance consciously a few times, and said, "Conscious breathing will bring the body and mind together. When the mind and body become one, we feel solid and become fully present for the tea. As we drink our tea, we should be completely aware that we are drinking the tea. When we truly meet the tea in the present moment, we feel alive. Only then are we really living our life." After a few attentive sips, we all experienced that drinking tea had become the most important thing at that moment. The light green tea and delicious melons complemented each other perfectly.

As we relaxed into the evening, we felt nourished—not only physically, but also spiritually. We did not feel stuffed but rather perfectly satisfied, content, fortunate, and peaceful.

Eating Mindfully Every Day

We may not all have an opportunity to dine with a Chef Sati, but we should try to follow his valuable example as much as possible.

At home, reserve a time for dinner. Turn off the TV; put away the newspapers, magazines, mail, and homework. If you are eating with others, work together to help prepare dinner. Each of you can help with washing the vegetables, cooking, or setting the table. When all the food is on the table, sit down and practice conscious breathing a few times to bring your body and mind together, and recover yourselves from a hard day's work. Be fully present for each other, and for the food in front of you.

After a few conscious breaths, look at each other with a gentle smile and acknowledge each other's presence. If you are eating alone, don't forget to smile to yourself. Breathing and smiling are so easy to do, yet their effects are very powerful in helping us and others to feel at ease. When we look at the food in such a moment of peace, the food becomes real and reveals our connection with it and with everything else. The extent to which we see our interrelationship with the food depends on the depth of our mindfulness practice. We may not always be able to see and taste the whole universe every time we eat, but we can do our best to eat as mindfully as possible.

When we look at our food on the table, it is helpful to name each dish: "pea soup," "salad," and so on. Calling something by its name helps us touch it deeply and see its true nature. And mindfulness reveals to us the presence or absence of toxins in each dish so that we can stop eating something that is not good for us. Children enjoy naming and recognizing foods when we show them how.

Being with our family and friends to enjoy food is precious. Many people are hungry and without family. When we eat in mindfulness, we generate compassion in our heart for them. With compassion and understanding, we can strengthen our commitment to helping nourish the hungry and lonely people around us. Mindful eating is a good educau practice this way for some time, you will find that you will eat i refully, and your practice of mindful eating will be an example for others. It is an art to eat in a way that brings mindfulness into our life.

The Seven Practices of a Mindful Eater

One way to incorporate mindfulness into your meals is to simply use the breath. Before eating, make a practice of pausing. Breathe in and out a few times so that you can be one with the food you are about to eat. Mindful eating takes dedicated practice, and there are seven practices that you can develop to help you eat mindfully for good health.

1. HONOR THE FOOD. Start the meal with the five contemplations, or with whatever traditional grace or prayer you prefer to use to express your gratitude.

The Five Contemplations

1. This food is the gift of the whole universe: the earth, the sky, numerous living beings, and much hard, loving work.
2. May we eat with mindfulness and gratitude so as to be worthy to receive it.
3. May we recognize and transform our unwholesome mental formations, especially our greed, and learn to eat with moderation.
4. May we keep our compassion alive by eating in such a way that we reduce the suffering of living beings, preserve our planet, and reverse the process of global warming.
5. We accept this food so that we may nurture our sisterhood and brotherhood, strengthen our community, and nourish our ideal of serving all living beings.

If you are eating with others, steer mealtime conversations toward the food: acknowledge the local farmer who grew your lettuce and tomatoes, thank the person who prepared the salad; or talk about other topics that help nourish your gratitude and connection to your food and each other. Refrain from hashing over work or the latest atrocities in the news. Refrain from arguing. This can help you make sure that you are chewing only your food, not your frustrations. In Vietnam it is a custom to never chastise anyone while they are eating, so as not to disturb their eating and digestion. We can learn from this very commonsense wisdom. Eating in this way, we have the opportunity to sit with people we love and to savor precious food, something that is often scarce for many people in the world.

At all Plum Village practice centers, we eat our meals in silence during the first twenty minutes of the meal so that we are fully immersed in the experience of eating. We encourage you to experiment with a silent meal at home—even just a silent cup of tea. But you do not need to eat every meal in silence to become a more mindful eater. You can start by simply unplugging from daily distractions during mealtime: turn off the television, the laptop, the cell phone, so there is no watching, no surfing, no texting.

2. ENGAGE ALL SIX SENSES. As you serve and eat your meal, notice the sounds, colors, smells, and textures as well as your mind's response to them, not just the taste. When you put the first bite of food in your mouth, pause briefly before chewing and notice its taste as though it was the first time you had ever tasted it. With more practice in engaging all of your senses, you may notice that your tastes change, increasing your enjoyment of what you may once have perceived as "boring" health foods.
3. SERVE IN MODEST PORTIONS. Moderation is an essential component of mindful eating. Not only does making a conscious

effort to choose smaller portions help you avoid overeating and weight gain; it is also less wasteful of your household food budget and our planet's resources. Using a small dinner plate, no larger than nine inches across, and filling it only once can help you eat more moderately.

4. SAVOR SMALL BITES, AND CHEW THOROUGHLY. Consciously choosing smaller bites and chewing them well can help you slow down your meal as well as allow you to fully experience the taste of your food. It can also help improve your digestion, since the process of breaking down our foods begins with enzymes in the mouth. Chew each bite until the food is liquefied in your mouth; that may be twenty to forty times, depending on what you are eating. Chewing well allows your tongue and palate to taste the food better. Once you have swallowed this bite, you will still be able to savor the wonderful taste that the food offers you.
5. EAT SLOWLY TO AVOID OVEREATING. Eating slowly may help you notice when you are feeling pleasantly satisfied so that you can stop before you have eaten too much. There is a difference between feeling that you have had just about enough to eat and feeling as though you have eaten *all* that you can possibly eat. Mindful eaters practice the former so that they are not overtaxing their bodies—or overtaxing the planet's resources—by consuming more food than they need. In Chinese medicine, it is recommended to eat only until you are 80 percent full and never to "top off your tummy," because this weakens the digestive power of your stomach and intestines, putting too much stress on them over the long haul. There is ongoing scientific research on the effects of caloric restriction on longevity, though the results are far from conclusive in humans.[62] Of course, avoiding overeating is half of the secret to weight control. (Getting enough activity is the other half, and we discuss this more in chapter 6.)

One way to slow down is to consciously put your eating utensils down in between bites. Be aware of your body as you eat. When we eat mindfully, we are relaxed and calm. There is no rush to attend to other tasks; there is no hurry. There is only the present moment. To help you practice this, make sure to allow enough time to enjoy the meal. If your mealtime is short—for example, during your lunch break at work—plan on a smaller meal rather than cramming down a large meal quickly.

6. DON'T SKIP MEALS. Skipping meals can make it harder to make mindful choices. When hunger consumes us, the strong forces of habit energy may lead us to grab whatever foods are close at hand—be they from a vending machine or a fast-food restaurant—and these foods may not further our healthy-eating or weight-loss goals. So-called *grazing*—moving from one food to another, a few bites of this, a few bites of that, without ever sitting down to a regular meal—can also work against your healthy-weight goals, because you may consume more food than you realize without ever feeling truly satisfied. So give yourself the opportunity to make mindful choices throughout the day; plan regular meals and, if it suits you, healthy snacks in between. It is also good to eat your meals at the same time each day, to help your body settle into a consistent rhythm. And give yourself enough time to fully savor your food so that you are aware of all the sensory delights your meals have to offer.
7. EAT A PLANT-BASED DIET, FOR YOUR HEALTH AND FOR THE PLANET. When mindful eaters look deeply at the meal they are about to eat, they see far beyond the rim of the plate. They see the dangerous toll that eating some types of animal foods can take on their bodies—the higher risks of colon cancer from red meat and processed meats, for example, or the higher risk of heart disease from the saturated fat found in meat and dairy products. And they see the equally dangerous and de-

structive toll that meat production and dairy farming take on our environment. Researchers at the University of Chicago estimate that, when it's all added up, the average American could do more to reduce global warming emissions by going vegetarian than by switching from a Camry to a Prius.[63] Even just switching from red meat and dairy to poultry or eggs for one day a week could have a measurable impact on global warming—and a bigger environmental impact than choosing locally sourced foods.[64]

Watch Out for Mindless Traps

We have talked about what to eat and how to enjoy our meals. We all would like to eat healthfully, but we also have our own internal knots, the strong habit energies that distract us from practicing mindfulness. People are more likely to eat healthfully if they believe that they are capable of doing so; if they believe doing so will have health benefits; if they have support from family and friends; and if healthy eating is the norm for most of their family members or coworkers.[65] People are also more likely to eat well if their community or workplace makes it easier to do so—if, for example, they live near supermarkets or if their worksite cafeterias and vending machines offer healthy foods.[66]

Likewise, barriers to healthy eating can exist within the person or within the environment the person is immersed in. Not wanting to give up favorite foods, not liking the taste of healthy foods, not trusting experts' seemingly ever-changing dietary advice—all of this can hinder our efforts to be healthy. People who live in "food deserts" without easy access to a supermarket may find it harder to eat fresh fruits and vegetables, and people who live in neighborhoods surrounded by fast-food restaurants may find it all too easy to choose higher-calorie, less-nutritious foods.[67] Food prices may also be a barrier: lower-calorie, more-nutritious foods, such as fruits and vegetables and fish, are more costly per calorie than higher-calorie, less-nutritious foods, such as refined grains and sweets.[68] Agricultural policies, food labeling, and food advertising can influence which foods are available to us in a supermarket or restaurant, whether we are aware of their nutrient benefits (or detriments), and whether we decide to eat them.

We will discuss how to build social support for healthy eating, as well as what steps you can take to advocate for a healthier food environment, in the following chapters. Here we will focus on the personal perspective: what are some of the common food-related habits that may be getting in the way of eating healthfully and achieving a healthier weight—and what are some tips to change them? Once you identify these personal barriers to healthy eating, you can start to overcome them and begin to make every meal a healthier one.

Do You Skip Breakfast or Other Meals?

Fewer people in the United States are starting the day with breakfast,[69] and our increasing failure to "break the fast" every morning may be contributing to the obesity epidemic. Research suggests that people who skip breakfast tend to weigh more and gain more weight over time than people who eat breakfast.[70] The National Weight Control Registry gathered information on the breakfast habits of nearly three thousand people who had lost and kept off significant amounts of weight, and found that nearly 80 percent of them report eating breakfast every day, while only 4 percent report skipping breakfast.[71] Scientists are still teasing out the nature of the relationship between breakfast and weight. It's possible that eating breakfast may help curb our hunger later in the day—and, in turn, the number of calories we eat, especially if the breakfast includes protein or fiber, such as that found in whole grains and fruits.[72]

Dieting books give all sorts of advice about when to have your meals—eat six small meals a day, stick to three meals a day and eschew snacks, don't eat after 8 P.M., and so on. The truth is that there's no "perfect" meal pattern that fits everyone's lifestyle or that is guaranteed to promote weight loss. A good approach, however, is to spread your food intake throughout the day.[73] This may mean three meals (one of them breakfast) and a snack or two, or four "mini-meals."

If skipping breakfast or other meals has become a habit, consider these suggestions for how to change your habit:

- PREPARE YOUR BREAKFAST OR PACK YOUR LUNCH THE NIGHT BEFORE. If you find that you are skipping breakfast or lunch because you don't have enough time to prepare it in the morning, plan breakfasts and lunches that can be assembled before you go to sleep. For lunch, set aside some of your dinner leftovers in a container, add a whole-wheat roll and piece of fruit or some baby carrots, and put it in a sack in the fridge so it will be ready for you to grab on your way out the door in the morning.
- BROADEN YOUR IDEA OF BREAKFAST. Some people skip breakfast simply because they do not enjoy traditional U.S. breakfast foods such as cold cereal and milk or eggs and toast. There's no reason to have such a limited palate. Try whole-grain hot cereals, combining whole grains and seeds such as sesame seeds, whole oats, whole rye, whole barley, millet, quinoa, and whole buckwheat. For convenience, you can cook a batch that lasts for a few days. Or try an Asian breakfast, with tofu or fish, vegetables, and brown rice. Make a breakfast burrito with beans, salsa, and a stone-ground-corn tortilla. Even last night's leftovers can make a nourishing and satisfying breakfast.
- MAKE SURE YOU'VE GOT A GOOD APPETITE FOR BREAKFAST. Overeating at night may make you less hungry for breakfast in the morning. If you're a nighttime snacker who has little appetite in the morning, you may want to try curbing your nighttime snacking to see if that improves your appetite for breakfast the next day. In extreme cases, eating at night may be characterized as "night eating syndrome," which we discuss later in this chapter.

Do You Speed-Eat?

It has become common advice for dieters: "eat slowly and chew your food well." And it certainly makes intuitive sense. The theory, popularized nearly forty years ago,[74] is that it takes twenty minutes for our brains to register that our stomachs are full, and that when we eat too quickly, we speed through the physical and hormonal "stop" signs and overeat. Eating more slowly may also give us more pleasure from our food as we take time to savor every bite. Many, many researchers have sought to test this notion, sometimes with conflicting results,[75] but recent support for the theory comes from a small study at the University of Rhode Island.[76] Researchers asked thirty women to eat a meal rapidly and, several days later, to eat a meal slowly (or vice versa), and measured the amount of food they ate and how satisfied they felt at the end of each meal. When the women ate a meal slowly, they consumed fewer calories and drank more water than when they ate a meal rapidly. After the slow meal, they reported greater satiety than they did after the fast meal. So although they ate *less* than at the faster meal, they felt more satisfied and full. It's interesting to note that Japanese studies on eating speed and weight control have found that people who say they eat rapidly weigh more and are more likely to be obese than people who say that they eat slowly.[77]

If speed-eating has become a habit, consider these tips to slow down and savor your food:

- MAKE YOUR FIRST BITE—AND EVERY BITE—A MINDFUL BITE. Remember the apple meditation in chapter 2, where we asked you to pay attention—to feel the weight of the apple in your hand, notice its color and smell, and think about all the natural and human forces that aligned to bring it from the earth to your hand? We suggest that you do the same before you begin eating all of your meals.

- TAKE SMALL BITES, CHEW YOUR FOOD THOROUGHLY, AND PUT DOWN YOUR UTENSILS IN BETWEEN EACH BITE. This has also been standard advice for people seeking to lose weight. And the University of Rhode Island study, which instructed participants to use these techniques during the slow meal, showed that taking these steps can help people slow down. Study participants' slow meals lasted, on average, nearly thirty minutes—about twenty minutes longer than their fast meals. Using smaller spoons and forks or using chopsticks (especially if you are not accustomed to using chopsticks) may help you train yourself to take smaller bites.

Do You Mindlessly Overeat Large Portions?

Often people overeat without being aware that they are overeating. They eat too much because they are snacking out of a supersize bag of chips, they have been served a heaping plate of food, they are watching television while they eat or because of any number of external cues that have nothing to do with hunger. Dr. Brian Wansink of Cornell University calls this type of eating "mindless eating," and he and other researchers have demonstrated the many ways that our environment can trigger us to simply eat too much.[78] Certainly there is ample opportunity to overeat: over the past few decades, portions have grown to gargantuan sizes in restaurants, grocery stores, and homes—even in the all-American classic cookbook *Joy of Cooking.*[79] These large portions lead us to unconsciously redefine what a "normal" portion size is, and also make it harder to estimate just how much we are eating.[80] Having tempting foods in sight, being distracted by a magazine, being offered a variety of foods—all can lead to mindless overeating.[81]

If you have gotten into the habit of eating with your eyes rather than your stomach, consider these tips to help you mindfully monitor your portion sizes:[82]

- USE SMALLER PLATES AND SERVING UTENSILS. Downsizing your plates, bowls, and serving spoons may help you downsize your portions.
- AVOID DISTRACTIONS WHILE EATING. It makes intuitive sense that watching television during meals could lead you to pay less attention to what you are eating and how full you are—and in turn contribute to mindless overeating. Other distractions—a movie, a social gathering, workplace demands—could have a similar effect. So separate eating from watching television or other activities. When you are engaged in a pleasantly distracting activity that revolves around food—such as a dinner party—recognize that you need to pay greater attention to what you put on your plate and in your mouth. Periodically take regular, mindful breaths throughout the meal to help you remember to come back to your body and check in with your stomach. Relaxing your whole body with your mindful breathing as you eat is a good way to help you stay in tune with yourself and your meal and to avoid overeating.
- ASK YOURSELF, "AM I SURE IT'S HEALTHY?" Our belief that a product is healthy or that a restaurant serves healthy food—what Wansink and colleagues call a "health halo"—can lead us astray when it comes to choosing moderate portions. People were more likely to underestimate the number of calories in a Subway meal than in a McDonald's meal, perhaps because of the sandwich chain's heavy marketing of itself as a "healthy" option, with its well-known advertisements featuring a famous customer's weight loss.[83] Before you place your order, stop, take a deep breath, and don't lose your mindful outlook just because a food is pitched as "healthy." Healthy foods have calories, and keeping calories in check is the key to weight loss.

Do You Eat a Lot at Night?

It's a common complaint from people who are trying hard to eat healthfully: "I make great choices all day, but after dinner, I just can't stop snacking." For some people, night eating means mindlessly munching on a bag of chips while they watch television or dishing out a bowl of ice cream as a reward for a stressful day. For others, so-called night eating syndrome is a more serious disorder of their circadian rhythms, one in which they consume at least 25 percent of their daily calories after dinner or wake up in the middle of the night to eat, leaving them with little or no appetite for breakfast.[84] Although night eating syndrome is not technically classified as an eating disorder just yet, it is estimated that anywhere from 6 to 16 percent of people enrolled in weight-reduction programs suffer from it[85] and that it can interfere with attempts to lose weight.[86] You may not have full-blown night eating syndrome (if you believe you do, consult with a professional), but if you find that nighttime has become synonymous with unhealthy snack time, consider these tips:

- FIND SOMETHING ELSE TO DO WITH YOUR HANDS. Knit, make a scrapbook, read a book out loud to your children, play chess—do something to keep your hands busy and your mind engaged so that you will be less tempted to turn to food.
- TURN OFF THE TELEVISION. As we will discuss, people tend to eat mindlessly when they watch television, and they tend to eat *what* they see in the commercials—mostly unhealthy snacks. Turning off the television and finding another nighttime activity can help you break the "eat-and-watch" habit.
- BE AWARE OF WHERE YOU WORK. Checking your e-mail at the kitchen table after dinner can mean that you are too close to tempting snacks. Move to a different room. If you must work in the kitchen, keep the most tempting foods out of sight.[87]
- REDUCE YOUR STRESS. There's evidence that stress may trigger night eating syndrome and that lowering stress may help

curb it: night eaters in a small one-week study who listened to a nightly twenty-minute muscle-relaxation tape before going to sleep reported less nighttime hunger and eating, and lower stress, than night eaters who did not de-stress.[88] This is a good time to try out the meditations offered throughout this book and on the companion Web site for this book. Or try some other relaxing activities, such as listening to soft music, reading a book, or taking a bath.

- GO TO SLEEP EARLIER. As we discussed in chapter 1, people who get less sleep tend to weigh more than people who get a good night's rest, and one reason may simply be that staying up later gives people more hours in the day to eat. Getting enough sleep may also help curb hunger during the day.
- IF YOU DO SNACK IN THE EVENING, CHOOSE FRESH FRUITS AND VEGETABLES. If you mindfully decide to answer your body's call for nighttime munchies, do it with fresh vegetables or fruits. They come packed with fiber, they fill you up quickly, and most don't have that many calories.

Do You Often Eat Fast Food or Restaurant Meals?

Americans spend more than 40 percent of their food budgets on food away from home,[89] and research has found that meals prepared outside the home tend to be less healthful than what we cook up in our own kitchens.[90] The best thing to do when you are trying to lose weight is to cut out fast food completely. Even the "healthy" choices aren't usually that healthy. Consider these tips to make the healthiest choices when you are out at fast-food or regular restaurants:

- DO YOUR RESEARCH AHEAD OF TIME. Many restaurants now have nutrition information available online, and it may surprise you. At some fast-food restaurants, for example, a small sandwich may be lower in calories than a large salad loaded with cheese and toppings.

- ORDER SMALLER PORTIONS. When we eat at a restaurant, once the food appears on our plates, chances are good that we will eat it—and the larger the portions, the more we will eat. So order a small appetizer instead of an entrée, share an entrée with your dining companion, or ask the waiter to bring you a "to go" container with your meal so that you can put half of your meal in the container before you have a chance to overindulge. And skip the extras—bread and butter, unlimited sugary fountain drinks—to save your appetite and calories for more nutritious and satisfying foods.
- DON'T BUY THE FAST-FOOD "MEALS." Going à la carte at a fast-food restaurant may cost a little bit more, even for buying less food, but who really needs to be tempted by the fifteen hundred calories in a complete meal with large fries, big burger, and thirty ounces of sugary soda? You'll be surprised at how filling a smaller sandwich and smaller fries can be, especially if you eat slowly and savor them.
- ORDER COFFEE OR TEA AFTER THE MEAL RATHER THAN DESSERT. Wansink notes that the relaxing atmosphere of a candlelit restaurant can make a meal more pleasant but can also lead people to spend a longer time at the table—and to eat more.[91] His suggestion: linger over a cup of coffee rather than a high-calorie dessert.

Do You Lack the Time to Prepare Healthy Meals?

Time pressures from work and family can make us all feel strapped, and lack of time for meal preparation is seen as a barrier to healthy eating.[92] But with careful planning, healthy eating does not need to take any more time than going out for takeout or popping a frozen pizza into the oven. Consider these tips:

- DIVVY UP THE DINNERTIME DUTIES. Involve family members or roommates in meal preparation; even young children can help. If you live alone, consider creating a healthy lunch or dinner swap with four of your friends or colleagues at work: Assign everyone to make a quintuple-size batch of a healthy entrée over the weekend, and divide the batches into portable containers. On Monday, bring them in and swap, and you will each have enough lunches or dinners for the week.
- BE REALLY CAREFUL ABOUT CONVENIENCE FOODS. You may turn to convenience foods to shave some time off meal preparation, but they can have their drawbacks. Some processed foods are loaded with added salt or sugar or are high in unhealthy fats. Look for frozen entrées that have less than 300 milligrams sodium, less than two grams of saturated fat, and zero grams trans fat and that offer at least a few grams of fiber per serving; add fresh fruit or a salad to round out the meal. Or make your own convenience "health" foods: cook up a large batch of whole grains, dried beans, or roasted vegetables on Sunday and use them at meals throughout the week or freeze them in small portions so that you can grab one for an easy lunch or dinner.

Do You Eat More on the Weekends Than During the Week?

Weekends are a time for relaxing, socializing, and, for many of us, overeating—and these small indulgences may lead dieters to gain weight or slow their weight loss. The National Weight Control Registry has found that people who stick to the same eating patterns on the weekends, on holidays, or on vacations as they do during the week tend to be more successful at maintaining their weight loss.[93]

If you find yourself falling away from your healthy eating intentions on the weekends, consider these suggestions to keep you on track:

- KEEP A FOOD DIARY. Research suggests that people who keep track of what they eat may be more successful at weight loss and at avoiding weight regain, since self-monitoring is a key element of self-regulation.[94] Jot down everything you eat and drink on your calendar, your smart phone, or one of several Web sites that offer free online food logs. You could even try taking a digital photo of what you eat: a picture is worth a thousand words, and there's evidence that taking pictures of what you eat may make you more aware of your food choices.[95] Don't agonize over the food-diary details; keeping the entries short may be just as effective for weight loss.[96]
- PLAN FOR MEALS WHEN YOU ARE ON THE GO. If you will be out running errands all day, pack healthy snacks in a cooler and take time for a mindful lunch break. If you will be eating lunch at a chain restaurant, scan the restaurant's Web site before you head out; most have nutrition information available that can help you pick the most healthful choices.
- MAKE YOUR SOCIAL TIME ACTIVE TIME, NOT EATING TIME. Rather than meeting a friend for a sugar-laden coffeehouse drink, meet for a brisk walk. Or go out dancing on a Saturday night instead of going out for dinner. (For more ideas on how to get active, see chapter 6.)

Do You Eat When You Are Angry, Sad, Bored, or Stressed?

The relationship between food and emotions is a complicated one. At one end of the spectrum is someone who has a stressful day at work and finds comfort in a candy bar on the way home. At the other end is someone who struggles with anorexia, bulimia, binge eating, or another type of disordered eating. Stress as well as several emotions may prompt someone to turn to food for comfort—among them anger, anxiety, boredom, loneliness, and sadness.[97] There are a variety of techniques for changing your behavior or changing your thought process to cope with emotional eating.[98] If you are concerned that emotional eating may be derailing your attempts at healthy eating, consider these tips to help you avoid using food to manage your feelings:

- USE MINDFULNESS TO RECOGNIZE THE DIFFERENCE BETWEEN PHYSICAL AND EMOTIONAL HUNGER. Before going to the refrigerator or heading to the snack-food aisle of the supermarket, stop, take a slow, deep in-breath, and on the out-breath ask yourself this question: am I truly hungry, or do I crave these comfort foods to ease my stress or relieve another emotion? If you keep a food diary, jotting down your mood and your level of hunger when you eat may help you identify if and when your emotions are leading you to overeat.
- SEEK ALTERNATIVE WAYS TO COPE WITH STRESS AND EMOTIONS. Walking, yoga, mindfulness meditation, singing along with your MP3 player, gardening, taking an herb-scented bath, having a phone conversation with a friend—these are just some of the many activities you can choose to help ease stress and provide a food-free alternative for channeling your emotions.
- KEEP TEMPTING FOODS AWAY FROM YOUR HIGH-STRESS ZONES. If you are under a lot of pressure at work, replace the candy jar on your desk with a squeeze stress ball or a desktop Zen fountain.
- CONSULT WITH A PROFESSIONAL WHO SPECIALIZES IN EMOTIONAL EATING. If you can't curtail your emotional eating on your own, you may want to find a psychiatrist or therapist who specializes in food-mood issues. Your doctor or your employee-assistance program at work can likely provide a referral. The National Eating Disorders Association has a toll-free information number as well as a directory of providers and sup-

port groups on its Web site: http://www.nationaleatingdisorders.org/get-help-today/.

Translating Knowledge to Action: Your *In*Eating Strategy

We have given you a menu of options for making healthy changes to your daily diet. We have reviewed the best food choices for health, the common barriers to healthy eating, and, so that you can truly savor your food, the connection between mindfulness and eating. Now it is time to put it all together and create a practical strategy that allows you to enjoy mindfulness at every meal—and to move closer to your goal of achieving a healthier weight. We call it your *in*Eating strategy, where the *in* means being in the present moment. With your *in*Eating strategy, you will be able to set goals for mindful, healthy eating and mindfully avoiding overeating, figure out ways around any barriers that might keep you from reaching your goals, and lay out the steps you will take to reach your goals. This *in*Eating strategy will be incorporated into your comprehensive Mindful Living Plan in chapter 7.

As we have discussed, mindful eating encompasses what we eat as well as how we eat—habits that are deeply ingrained and that require dedicated effort to change. While there are many possible changes you can make on either of these dimensions of mindful eating, only you can decide which changes are the most important to you and the most doable. To help you identify your *in*Eating goals, what will help you reach them, and the barriers you may face, reflect on the series of questions that follow. Journaling may help you as you consider your responses and chart your strategy.

Use your reflections to choose the goals that are most important and doable for you, and create your own *in*Eating practice, like the example in table 7.1. Once you are successful at changing one habit, build on your success to make other healthy changes. Keeping a log of your *in*Eating progress may help you better achieve your goals.

Like all action plans, the *in*Eating strategy is not meant to be a static document. As you gain experience with healthy, mindful eating and work through barriers, your goals and tips for overcoming barriers may change. Bear in mind the impermanent nature of all we do. Don't be afraid to adjust things, as long as you make sure you stay committed to and working toward your ultimate goal: choosing healthy, delicious foods at every meal, making every bite a mindful bite, and losing weight and keeping it off.

Why do you want to eat more healthfully and mindfully?

Think about the reasons you want to make healthier food choices and choose smaller portion sizes. And think about all the reasons you want to bring mindfulness to your meals. They can cut across all parts of your life. It would be useful to write these reasons down in a journal so that you can reflect on them later.

Examples: *I want to feel better about myself. I want to lose weight. I want to lower my bad cholesterol. I want to lower my risk of diabetes. I want to lower the carbon footprint of my food choices. I want to slow down so I can truly savor my food.*

What's bad about eating foods that are unhealthy for you or for the planet, eating too much, or eating mindlessly?

Think about the downsides of choosing unhealthy foods—for your own health and for the environment—and the downsides of eating more food than your body needs for nourishment. Think about the downsides of not paying attention to the act of eating or of ignoring the full implications of your food choices. Again, they can cut across all parts of your life.

Examples: *I'll stay heavy. I won't feel good about myself. I'll see my cholesterol go up. I'll be wasting money and the planet's resources by eating more food than I need.*

What healthy foods do you like? What healthy foods would you be willing to try? What mindful eating practices would you be willing to try?

Think about the healthy foods that give you joy. Think about the healthy foods that you do not currently eat but could consider adding to your meals. Write down those foods and the reasons that you want to make them a part of your meals. Then, think about all the practices of mindful eaters, which of those practices you would like to incorporate, and why.

Food examples: *Dark leafy greens, since they will give my bones calcium and vitamin K. Plant proteins like walnuts and lentils, since they are better for the planet and have beneficial nutrients for my body. Brightly colored fruits such as strawberries and blueberries, because their natural sweetness will satisfy my sweet tooth without sending my blood sugar soaring.*

Mindful eating practice examples: *Focusing on my food will give me greater enjoyment during my meals. Choosing smaller portions will be better for the planet and will also help me curb my daily calorie intake.*

What are some of the unhealthy foods that you could trade for healthier foods? What are some of the unmindful habits that you could most easily leave behind?

Think about the foods you eat that your body does not require for nourishment and that may actually harm your health—sugary drinks, refined grains, processed meats, salty snacks. Write down these foods and how you could trade them for healthier foods. Think about the habits you have that do not serve your desire to be fully present during the act of eating.

Food trade examples: *I can choose unsweetened iced tea instead of soda pop. I can snack on crunchy vegetables instead of chips. I can buy whole-wheat pasta instead of white pasta. I can drizzle olive oil on my vegetables instead of butter.*

Mindful eating habit examples: *I can have a silent cup of tea in the morning instead of spending my break calling a friend on my cell phone. I can tell my supervisor that I will be taking a full half hour for lunch instead of scarfing down lunch at my desk, and then make up that time later.*

In what meal during the day would it be easiest for you to trade healthy foods for unhealthy foods? In what meal during the day would it be easiest for you to incorporate one or more of the seven practices of a mindful eater?

Your ultimate goal is to make every meal a healthy, mindful meal. But for some people, it can be overwhelming to attempt to change everything all at once. You may find it easier to pick one meal a day to start—perhaps the meal in which you have the most control over your food choices, the most time to eat, or the fewest distractions.

Food examples: *I can make healthy food trades at:*

- *breakfast, since it's the one meal I eat at home every day*
- *lunch, because there's a great salad bar in the cafeteria*
- *the food court, because the burrito shop offers brown rice*
- *the gym, because I can fill my water bottle instead of buying an energy drink*
- *Sunday dinner, because I shop at the farm stand in the morning and cook a vegetarian dinner for the whole family*
- *evening snack, since I can eat fruit instead of cookies*

Mindful eating practice examples: *I can incorporate a new mindful eating practice at:*

- *breakfast, since I usually eat alone*
- *lunch, since there is a great vegetarian restaurant in the food court where I can use chopsticks and experiment with taking smaller bites*

- *dinner, since I can use a smaller plate to help myself choose smaller portions*

What are two or three obstacles that can make it hard for you to choose healthy foods and moderate portions? What are obstacles that could make it hard for you to eat more mindfully? What are a few ways around each of these?

Finding your way around obstacles that get in the way of healthy eating is a necessary and ongoing part of the process, and this is so for everyone, from a professional nutritionist to a person just starting out to make healthy choices. Think about some of the key things that keep you from making healthy food choices or lead you to overeat. They may be part of the list of barriers and habits we went over earlier in this chapter, or they may be something else entirely. Once you have the top two or three, write them down, and then come up with ways you can get around them. This will be your go-to list when things get tough and you are looking for reasons to return to your old ways of eating.

Examples:

Obstacle: I don't have enough time to eat breakfast.

My solution: I'll get my cereal and raisins ready on the counter the night before. Or I'll get up fifteen minutes earlier so that I have time to eat breakfast in the work cafeteria before going to my desk.

Obstacle: I snack mindlessly while I watch television late into the night.

My solution: I can track my TV time and make sure it's less than one hour at night. I can make a pact with a friend to call each other with a turn-off-the-TV reminder at a preappointed time to encourage each other to get a good night's rest; to help us consume programs more mindfully, my friend and I can also share about the program we just watched and discuss which seeds it watered in us. Before going to the fridge at night or opening a bag of chips, I can ask myself whether I am truly hungry. I can go to sleep earlier.

The Bottom Line

Mindful eating is a way to incorporate mindfulness into one of the most fundamental activities of our existence. It is a way to nourish our bodies and our minds. It is a way to help us achieve a healthier weight, and a way to appreciate the relationship between the food on our table, our health, and the health of the planet. It is a way to grow our compassion for all living beings and imbue reverence for life into every bite.

For something so rich in positives, though, it's not the easiest of steps for us to take. Living as we all are in a society that spends so much time and money to promote unhealthy foods and mindless eating, and to limit access to healthy foods, it takes a dedicated, mindful effort to be able to focus on and choose the foods that are best for our bodies and best for our planet. The steps described in this chapter give you the tools you need to listen to your body, to live in the moment, and to become a truly mindful, healthy eater.

CHAPTER 6

Mindful Moving

BEING ACTIVE IS ONE of life's miracles. It lets us play with our children, climb a mountain peak, or take a relaxing walk around the neighborhood with our friends and family. It also helps us focus our minds and get in touch with our senses, and it is one of the best and easiest ways to practice mindfulness in our daily lives. The systematic and deliberate nature of exercise—whether it's walking or running or doing yoga—grounds us in the moment and connects us with our thoughts and what we're seeing, hearing, and feeling. Physical activity can become an anchor of mindfulness in our day, and it's also one of best paths toward weight loss.

THAT'S THE GOOD NEWS. The hard news, which we all know, is that it can be difficult to get the exercise we need to keep ourselves healthy and to keep our weight in check. It can be so much easier to stay on the couch than to lace up our shoes and get out the door for a walk. Our minds can easily conjure up excuses, and if this describes you, take heart that you are not alone: over half the adults in the United States don't get the minimum amount of physical activity that they should.[1]

But don't despair. Our bodies yearn to be active. It's what they were made for. All we have to do is unleash that natural state of activity that we all have in us and that has been locked away by the automated, fast-paced, yet sedentary modern world in which we live. For to not be active, to not explore our physical selves and our connection with our senses, is to deny ourselves a treasure of well-being and a chance for personal transcendence.

As we have talked about in previous chapters, we live in a frenzied world that is often so filled with outside stimuli that most of our day is spent disconnected from our inner selves. We get lost in e-mail, the Internet, BlackBerries, television, or jumping from one social occasion to the next. Exercise, especially in mindfulness, gives you a break from all this and puts you back in touch with yourself and your senses. Let's call it *mindful moving.*

When we consume food, we are actually taking in energy and storing it in our body physiologically. Much of the energy is to support the essential physiological and chemical processes in the body, but an important part of it is spent on moving our body physically, mostly through the hard work of our muscles. When we breathe, the medium is air; when we eat, the medium is food; and when we move, the medium is energy. When we walk, our muscles burn the stored energy and convert it into kinetic energy for our limbs, allowing us to move. This energy exchange is a miracle of life. When we look deeply into the nature of our physical energy, we can see that its sources are the sun, the rain, the air, the earth, and our food. Like conscious breathing and conscious eating, conscious moving can also lead us to the realization that everything is dependent on everything else.

Mindful moving is not just exercise for the sake of exercise. It's an expression of our mindfulness practice that helps us touch the peace within, treasure what we have, and take real steps toward improving our health and well-being.

LET'S BEGIN THE MINDFUL-MOVING journey by taking a look at the sound science that backs the links between physical activity, health, and well-being. We will address questions such as: Why do I need to stay active? What are the benefits? How much physical activity do I need each day? What are the best types of physical activity for me?

Benefits of Exercise

When it comes to health and well-being, regular exercise is about as close to a magic potion as you can get. In the U.S. *2008 Physical Activity Guidelines for Americans,* the list of the benefits of exercise is so lengthy that it takes up an entire page (see table 6.1).[2] Evidence shows that regular activity lowers the risk for many chronic conditions, including diabetes, heart disease, high blood pressure, and unhealthy cholesterol, as well as osteoporosis and certain cancers. It's been shown to boost mood, improve quality of life, and help people better deal with life's stresses. It also helps prevent weight gain and obesity and helps people maintain weight loss. To top it all off, it's also been shown to extend life span. Really, the only thing that boosts health more than exercise is not smoking.

In addition to the many physical benefits of exercise, studies have shown that exercise has a profound impact on brain functioning, even if the exercise doesn't begin until later in life. Dr. Kenneth Cooper—the father of aerobics—and his son Dr. Tyler Cooper write in their book *Start Strong, Finish Strong: Prescriptions for a Lifetime of Great Health* that individuals in their forties who walk briskly for three hours a week can "stop the shrinkage of brain areas that are related to memory and

Table 6.1 HEALTH BENEFITS ASSOCIATED WITH REGULAR PHYSICAL ACTIVITY

Children and Adolescents

Strong evidence

- Improved cardiorespiratory and muscular fitness
- Improved bone health
- Improved cardiovascular and metabolic health biomarkers
- Favorable body composition

Moderate evidence

- Reduced symptoms of depression

Adults and Older Adults

Strong evidence

- Lower risk of early death
- Lower risk of coronary heart disease
- Lower risk of stroke
- Lower risk of high blood pressure
- Lower risk of adverse blood lipid profile
- Lower risk of type 2 diabetes
- Lower risk of metabolic syndrome
- Lower risk of colon cancer
- Lower risk of breast cancer
- Prevention of weight gain
- Weight loss, particularly when combined with reduced calorie intake
- Improved cardiorespiratory and muscular fitness
- Prevention of falls
- Reduced depression
- Better cognitive function (for older adults)

Moderate to strong evidence

- Better functional health (for older adults)
- Reduced abdominal obesity

Moderate evidence

- Lower risk of hip fracture
- Lower risk of lung cancer
- Lower risk of endometrial cancer
- Weight maintenance after weight loss
- Increased bone density
- Improved sleep quality

higher cognition."[3] They also explain that exercise is associated with improvements in verbal memory and attention.

Dr. John Ratey, author of the book *Spark: The Revolutionary New Science of Exercise and the Brain,* also talks about the many mental and emotional benefits linked to regular physical activity.[4] According to Ratey, exercise is key to enlarging brain cells, fostering learning, and decreasing stress, anxiety, and depression. He also states that exercise can help manage attention deficit disorders and addictive behaviors as well as help women deal with a lot of the unpleasant symptoms of menopause. Neural plasticity is an area of active research focusing on ways to change the neural activities and promote structural changes in the brain.[5] In his book *The Mindful Brain,* psychiatrist Dr. Daniel Siegel indicates that both aerobic exercise and mindfulness practice can promote neural plasticity.[6]

Most important for the topic at hand, regular activity is an essential part of any weight-loss plan. A 2009 scientific review of over forty weight-loss studies found that regular activity by itself could lead to weight loss if people do not match their energy expenditure by eating more.[7] When combined with less caloric intake, the losses were even greater. And as you might expect, the harder and longer people worked out, the more weight they lost.

So How Active Should I Be?

This is the common and challenging question for most of us with busy schedules and no plans to try out for the Olympic marathon team. Overall, the quick answer is that any activity is better than none; and in general, the more active you are, the better. But you don't have to quit your day job and devote yourself solely to the gym to see real benefits. According to Dr. Kenneth Cooper, going from no regular activity to walking two miles at a brisk pace five times per week can dramatically lower the risk of high blood pressure and heart disease while helping to improve mental and emotional health and manage stress.[8] Increase the intensity by choosing more vigorous activities, such as jogging, step aerobics, or cycling, and you can get the same benefits in just twenty minutes of exercise three days a week.

These levels of activity generally align with the U.S. federal guidelines for optimizing health benefits with exercise. It is currently recommended that adults get two and a half hours a week of moderate-intensity aerobic activity (such as brisk walking, water aerobics, or ballroom dancing) or one hour and fifteen minutes of vigorous physical activity (such as jogging or running, swimming laps, or jumping rope; see table 6.2).[9] Adding more activity to the daily routine will have bigger health payoffs. For those looking for more benefit, the guidelines recommend five hours per week of moderate-intensity aerobic activity or two and a half hours per week of vigorous-intensity exercise. Plus, they strongly encourage activities that strengthen our muscles, such as weight training, push-ups, and sit-ups at least two days a week.[10]

Don't Forget Strength and Flexibility

While the main activities focused on weight control are things like walking and jogging, it's important not to forget strength and flexibility exercises as well. And this is especially important as we age. Together, strength and stretching exercises build muscle, improve agility and balance, and help you avoid injury when you're doing your aerobic activities. You should do strength and stretching exercises, in addition to aerobic activities (see table 6.3).

Like most types of activities, you can make strength training as complex or as simple as you'd like. At the most basic, you want to do it two or three times per week, with rest days between each session. Ideally, the exercises you do should work all the major parts of your body. And whichever exercises you're doing, you want to do several repetitions of each. Typically, trainers recommend eight to twelve repetitions for each exercise, but for some individuals, as few as three repetitions or as many as twenty could be appropriate. Some trainers may recommend doing two or three sets of the exercises, but research suggests you can achieve similar benefits by doing just a single set of exercises—and you may be more likely to stick to a shorter, single-set session.[11]

If you join a gym, a trainer can get you started on an individualized routine. If you work out at home, many books can help you create a routine with items around the house. Or you can buy a resistance band or similar device and follow the simple routines that come with the instructions.

Flexibility training should be done more often than strength training. Do it most days of the week, even every day. And it doesn't really take too long. Always warm up a bit beforehand; then run through a series of stretches that cover all the major parts of your body. Again, many books or Internet sites can help guide you. *Mindful Movements: Ten Exercise for Well-Being,* by Thich Nhat Hanh, may be particularly helpful, as it combines activities that focus on strength, flexibility, agility, and mindfulness.[12]

If you're looking for something more formal or something you can do with a group outside the house, tai chi, yoga, Pilates, or an aerobics class that includes stretching as part of the warm-up or cool-down are good alternatives as well.[13]

Supershort bouts of exercise done throughout the day—say, multiple ten- to fifteen-minute exercise sessions—can be a good way to fit exercise into the daily routine, especially when you're just starting out. Though combining short bouts together may not be the best path toward weight loss, it may help prevent weight gain.[14] The important thing is simply to get moving regularly. If this means stringing together short bouts of activity throughout the day, then that's what you should do, and there's evidence that it has many benefits.[15]

Exercise Recommendations for Losing Weight

For people who are overweight, recommended amounts of activity reach a bit beyond those for people who are just looking to improve their health. The *2008 Physical Activity Guidelines for Americans* suggests that most people who want to lose weight need to get at least five hours of moderate physical activity each week.

This may sound like a lot, especially if you are not currently getting much exercise. Take heart: these levels don't need to be reached overnight. You should build up slowly, starting with an initial goal of two and a half hours each week and finishing at five hours or more each week. It can seem daunting at first, but building up slowly helps you get into a routine, and before you know it you'll feel like something is missing from your day when you don't get your exercise in.

Although there's some debate about whether exercise alone can achieve significant weight loss in people—with some researchers concluding that it can[16] and others concluding that it can't[17]—one thing is clear: the most effective way to lose weight is by combining increased activity with a healthy diet that cuts back on calories. The American College of Sports Medicine (ACSM) recommends that people trying to lose weight increase their activity and cut back on calories so that they create an overall calorie deficit of about five hundred to one thousand calories a day.[18] This would translate to losing one to two pounds a week. What does this mean to real people? It's really pretty simple. To create an overall calorie deficit of five hundred calories in a day, you can burn about three hundred calories more than usual by walking about three miles (one mile generally equals one hundred calories) and cut back on what you eat by about two hundred calories by skipping a bag of chips or a bottle of sugar-sweetened soda.

Of course, you can do other combinations. And to get up to one thousand calories a day deficit, you just increase both sides of the equation: get more activity and cut back on more calories. However you do it, it's important to be sensible. Weight loss isn't a sprint; it's a continuous journey. So while you need to make deliberate changes that make a difference, you also need make sure they are changes you can maintain over the long term. You don't want to start out your first week by going to the gym for three hours a day only to get burned out and not go back for two years. Take it slowly. Gradually increase your activity and shave off a few junk-food calories to achieve your sensible goal of a five-hundred-to-one-thousand-calorie deficit per day, and keep it there.[19] Though the pounds won't melt off like magic in the first few days, you can feel confident that they *will* drop off, and by doing things the right way, they'll be much more likely to stay off in the long term.

The ACSM guidelines encourage those trying to lose weight to strive for an initial goal of 150 minutes (2.5 hours) of moderate-intensity exercise (such as brisk walking, ballroom dancing, bicycling) each week or 20 minutes of vigorous-intensity exercise (such as jogging, aerobic dancing, singles tennis, jumping rope, or swimming laps) three times per week, which is similar to the recommendations of the *2008 Physical Activity Guidelines for Americans* for optimizing health benefits. After reaching and maintaining this amount of exercise, individuals should increase their energy expenditure to at least 200 to 300 minutes of exercise per week. This is the equivalent of 40 to 60 minutes a day in one or multiple segments, five days a week.

Keeping It Off

Of course, just as important as taking off the weight is keeping it off. And regular physical activity continues to be a key ingredient in weight maintenance. Both the *2008 Physical Activity Guidelines* and the ACSM guidelines recommend two to three hundred minutes of activity per week to keep weight off over the long term. Some evidence points to even more. Data from the National Weight Control Registry, which tracks nearly four thousand people who have lost significant amounts of weight, found that people are most successful at keeping the pounds off if they regularly got sixty to seventy-five minutes per day of moderate-intensity activity like brisk walking, or thirty-five to forty minutes per day of more vigorous activity such as jogging.[20] These are average amounts of activity, and there was quite a bit of variability in the amount of activity required; some people needed more activity to stay at a healthy weight, and some required less. It's very individualized,

Table 6.2 EXAMPLES OF DIFFERENT AEROBIC PHYSICAL ACTIVITIES AND INTENSITIES

Moderate Intensity

- Walking briskly (3 miles per hour or faster, but not race-walking)
- Water aerobics
- Bicycling slower than 10 miles per hour
- Tennis (doubles)
- Ballroom dancing
- General gardening

Vigorous Intensity

- Racewalking, jogging, or running
- Swimming laps
- Tennis (singles)
- Bicycling 10 miles per hour or faster
- Aerobic dancing
- Jumping rope
- Heavy gardening (continuous digging or hoeing, with heart rate increases)
- Hiking uphill or with a heavy backpack

Source: U.S. Department of Health and Human Services, *2008 Physical Activity Guidelines for Americans* (Washington, DC: U.S. Department of Health and Human Services, 2008).

and you'll need to determine this for yourself. If you're eating the same amount of food and your weight is creeping up, you may need to exercise more.

Television, Weight, and Health

For many of us, television has become a daily companion. It keeps us informed about happenings across town and around the globe. It alerts us about severe weather and other breaking events. And it's generally considered a source of entertainment and relaxation.

Unfortunately, a lot of us have taken TV watching to the extreme, letting it become a huge part of our daily lives. According to Nielsen Media, the average American watched four hours and forty-nine minutes of television each day during the 2008–2009 season, and current

Table 6.3 HOW ACTIVE SHOULD I BE?

	For Health Benefits	For Weight Loss and Health Benefits	For Keeping Weight Off and Health Benefits
Moderate-intensity aerobic activity	150 min./wk.	200–300 min./wk.	200–300 min./wk.
Strength training (weights, push-ups, power bands, etc.)	2–3 days/wk.	2–3 days/wk.	2–3 days/wk.
Flexibility training/ stretching	4–7 days/wk.	4–7 days/wk.	4–7 days/wk.

Sources: American College of Sports Medicine, *ACSM's Guidelines for Exercise Testing and Prescription* (Philadelphia: Lippincott Williams & Wilkins, 2006); J. E. Donnelly et al., American College of Sports Medicine position stand: Appropriate physical activity intervention strategies for weight loss and prevention of weight regain for adults, *Medicine & Science in Sports & Exercise* 41 (2009): 459–71; U.S. Department of Health and Human Services, *2008 Physical Activity Guidelines for Americans* (Washington, DC: U.S. Department of Health and Human Services, 2008).

trends suggest that this number will continue to rise each year.[21] Kids spend more time watching television and playing on computers each year than they do in school, and the weekly screen time for adults is getting close to the equivalent of a full-time job.

It's understandable how this happens. Companies spend billions of dollars to entice you to watch their shows and their ads so you will buy their products and then come back and watch even more ads. And they're all very good at their jobs. But as we let ourselves and our kids get caught up in this vicious cycle of watching and buying, our mental and physical health suffers.

As mentioned in chapter 1, time spent watching television ties directly to rates of obesity. Studies have shown that men who watched three or more hours of television per day, and women who watched four or more hours of television per day, were two times more likely to be obese than those who watched less than one hour a day.[22] The large-scale and long-running Nurses' Health Study has shown that the risk of obesity in women increased by nearly 25 percent for each two-hour block of time women spent watching TV each day.[23] Even those who were very physically active couldn't fully escape the obesity risk linked to large amounts of television viewing.

And the downsides of television viewing don't stop just at weight gain. It's an equal-opportunity hazard that has also been found to increase the risk of diabetes, as well as heart disease risk factors such as high blood glucose, high blood pressure, high triglycerides, and low HDL ("good") blood cholesterol.[24]

So how is it that television can be so bad for you, even in those who are physically active overall? There are several possible reasons. Time spent watching television is time you're not spending being active, even if it's just padding around the house doing various things. This means you burn fewer calories, which can lead to weight gain and a host of unhealthy risk factors. In addition, some researchers posit that television watching is strongly associated with mindless consumption of unhealthy high-calorie snacks and beverages, since many people tend to snack while watching television. A good example is a 2006 study that found that subjects who watched television ate significantly more food than did a similar group who listened to classical music.[25] Taking this idea a step further, researchers suspect that the food ads on television increase consumption not only while people are in front of the TV but also throughout the day. And there is some good evidence to back this up. One study showed that those who watched the most television were more likely to eat dinner at fast-food restaurants.[26] And another study showed that those who watched two or more hours of television per day had the highest intake of calories, while those who watched less than an hour had the lowest.[27]

The result of this sedentary lifestyle and an increased intake of often-unhealthy calories is a toxic mix that not only promotes weight gain but also increases risk factors for heart disease, diabetes, and stroke.

Because of all this, it's important to be mindful about how much television you watch. Most recommendations say that adults and children shouldn't get any more than two hours a day. And the closer you can get this figure to zero, the better. This can be hard to do for many of us, but this is where practicing mindfulness can be helpful in changing our television-viewing habits.

Television-Viewing Meditation

When you find yourself reaching for the television remote control, stop and take a few in-breaths and out-breaths.

> Breathing in, the remote control is in my hand.
> Breathing out, why am I watching television?

By pausing and breathing, you are breaking your ingrained autopilot habit of watching television whenever you are tired, bored, feeling restless, or wanting to relax. Breathing in and out mindfully helps you reconnect to your current state of feelings and thoughts. Being mindful helps you be in touch with what is really helpful to cope with the particular physical or emotional state you are in. You may then realize that television is not the antidote for your current state of body or mind. If you are tired, the images and sounds from television stimulate your senses more instead of helping you relax and rest. Just breathing in and out helps you realize that a wiser solution to your tiredness is to lie down in a comfortable position, close your eyes, and focus on your breath to allow your body to calm down and truly rest. Or if you are feeling blue, by being mindful you realize that it is better to call your good friend to chat about what is bothering you than allow the sights and sounds of television to camouflage an emotion that needs your care. We invite you to explore appendix D, which offers more than fifty creative alternatives to television watching, many of which involve physical activity.

Barriers and Challenges That Prevent Physical Activity

Now that we have covered some of the benefits of activity and how much we need it to stay healthy and keep our weight in check, it's time to figure out how to actually make that happen. For a few people, it's as easy as deciding to be more active, buying a good pair of walking shoes, and just getting out the door to a new, active life. For most of us, though, it's a bit harder than that. There are a lot of things in life that can keep us from getting the activity we need. They can be mental barriers, physical barriers, social barriers, or any manner of things that work against our getting out that door for a workout.

For our health and happiness we need to figure out a way around these barriers and carve out a permanent space in our lives for activity. Once you do this, before you know it, activity will become an indispensable part of your daily routine that keeps you on the path of fitness, good health, and mindfulness.

If you asked most people whether being more physically active would be good for their health and mental well-being, almost all would say yes, yet according to the latest data, more than half of all adults in the United States don't get enough physical activity, and 25 percent don't exercise at all.[28] There are a lot of reasons for this disconnect between knowledge and action, some that exist within ourselves and some that exist outside ourselves. With the rare exception, though, we can navigate around and over most such barriers and find our way to a more active life.

What makes active people active and sedentary people sedentary? Many studies have sought to answer that question, and the findings may not come as much of a surprise: People are more likely to be active if they believe they'll be successful at it—that they'll be able to do it regularly, they won't be too tired, they'll be able to do it without feeling bad or embarrassed.[29] People are more likely to be active if they believe they'll get benefits from it—that they'll feel better, they'll lose weight, they'll lower their risk of heart disease.[30] People are more likely to be active if their physical and social environment supports an active lifestyle—if they live in safe neighborhoods, have access to parks and walking paths, and have family and friends who encourage them.[31]

The flip sides of these positive forces that make it easier for people to be active are the barriers that can keep us from being active. Leading the list for most of us: lack of time. But myriad other obstacles can stand in the way of becoming more active, like fear of not meeting our exercise goals, fear of being injured, fear of being made fun of, not having enough money for the right equipment or gym membership, or simply disliking exercise.

These are all valid concerns. But with a little creativity and forethought, you don't need to give up on your active-living journey before it starts. How can you overcome these barriers? You simply need to make a plan. Figure out what you want to do and how you are going to get there. Even the person for whom it seems exercise comes as second nature has to work all the time to overcome barriers to staying active.

The cornerstone of any plan is commitment: commitment to being active, commitment to losing weight, and commitment to the plan itself. While this sounds a bit stark, it's really not. Our lives are dynamic entities: our schedules change, our attitudes change, our relationships develop. So any plan has to be flexible enough to work through all of these factors, and more. But your commitment to the goal of being more active—and the goal of *finding a way* to become more active—is essential. And it's easier than you think, although it doesn't happen overnight.

Be warned that whenever you try to adopt a new habit, it is natural to backslide a bit. If you find that you have fallen away from your exercise plan for a day, a week, even a month, do not judge yourself harshly as a failure. Just start again. Every day is a new day, an opportunity for a new beginning. You can begin anew every moment. Start small, with baby steps, and before you know it activity will be as much a part of your life as sleeping and brushing your teeth.

Now, let's address some of the more common barriers people have when it comes to being more active, and the steps you can take to overcome them.

"I Just Don't Have Time"

If there were an award given for the most common issue that keeps people from exercising, the hands-down winner would be: "I just don't have time." And it's certainly a valid issue. People have busy schedules—work, family, household chores, soccer games, school recitals, book clubs, and more. Time is precious and hard to come by. But this doesn't mean you can't carve out space for regular activity, especially when you think about the health boost you will get and how it will help you maintain your healthy weight. Try to make exercise as essential a daily routine as eating or sleeping. It may be tough to fit it in at first, but after a while the day just won't seem complete unless you've had your exercise. You'll miss the renewal and joy you get from exercising.

Consider the following ways to work around this barrier:

- WAKE UP EARLY. When the day gets rolling up to full steam, it can be hard to find the time to exercise. Wake up early and do yoga in your living room, head out the door for a brisk walk to enjoy the fresh morning air, or head to the gym before others in the house are up needing your help or company.
- ENERGIZE YOUR LUNCH. After mindfully eating your lunch, gather some friends and head for a walk before you have to head back to work or your chores. Even a fifteen-minute walk after lunch helps reenergize you during midday.
- GET YOUR WORKOUT CLOTHES AND SHOES TOGETHER THE NIGHT BEFORE. This way you'll be packed and ready to go—either in the morning for an early workout or later in the day when a workout window opens.

- FIT IN SOME EXERCISE HERE, SOME EXERCISE THERE. If you just can't fit in one solid block of exercise, try stringing together a few smaller bouts: a twenty-minute walk first thing in the morning; a twenty-five-minute walk at lunch; and a fifteen-minute walk home from the store after work. Together, you've reached sixty minutes for the day.
- FIND THE PHYSICAL ACTIVITIES THAT YOU ENJOY. Explore different types of physical activities—bicycling, dancing, Frisbee, hiking, table tennis, swimming, tai chi, rock climbing, golf, yoga, and more. Find what you really enjoy. It's easier to stick to a physical activity routine when you have fun doing it.

"I'm Too Tired"

Don't let the "tired" excuse drag you down. It's an easy one to use, and one we can all relate to. Our lives are filled with many, many responsibilities that take up a lot of our energy, and when we finally find a spare moment to ourselves we often want to put our feet up and relax rather than lace up our shoes and get out the door for a walk.

And, of course, there are times when we're just so tired that there's nothing we can do but crawl into bed when we have the chance. But when you think about it and examine how you are feeling right in the moment, you will come to realize that these times are very rare. In most cases, getting some exercise is just what we need to clear our minds of the day's problems, make us feel empowered, and give us a boost of energy to get us through the rest of the day.

Consider these ways to work around this barrier:

- EXERCISE DURING YOUR MOST ENERGIZED PART OF THE DAY. We all have parts of the day where we feel better than others. Some people are morning people. Some people are evening people. Some people like midday. Whichever you are, try to schedule your workout during that part of the day when you usually feel your best.

- TURN OFF THE TELEVISION. Nothing saps one's energy like television. Whether it's a good show or bad show, once you sit down in front of the television you are much less likely to get up and go get the exercise you need. So, before you decide to sit down and watch your favorite shows, get out the door for a brisk walk, ride your bike, or shoot a few baskets. Perhaps you will be so invigorated from your workout that you'll decide to do something else that you really need to do with your time instead of watching television.
- SCHEDULE WORKOUTS WITH FRIENDS. Nothing helps keep your workouts on track—even when you're tired—like support from your peers. Get a group of friends together, and set a regular time to meet them for a trip to the gym or a good jog. They'll keep you honest—not letting you off too often for being too tired to work out—and you'll be doing the same for them.
- KEEP AN EXERCISE LOG. Whether you do it online or in a notebook, keeping a log of your workouts can be a great way to motivate yourself and get yourself out the door. You'll be amazed at how easily you'll wave off your fatigue and just get out of the door when you're faced with entering zero minutes of workout time for a whole day.

"I Don't Want to Look Silly"

Almost anyone who's ever been to a gym has felt awkward compared with the fitness buffs at the weight machines, with their toned triceps and lean physiques. It's only natural, especially for many people who are overweight or don't feel comfortable with their bodies. But there's no reason to let these feelings stop you. Your health and your weight are important priorities, not what people may or may not be thinking about you. This is another opportunity to practice mindfulness by fully immersing yourself in the present moment without judgment. Focus on your body and the workout, the movement of your arms and limbs—not the people around you. And the more you exercise, the more time you spend at the gym or jogging on the bike path, the more comfortable you'll feel.

Consider these ways to work around this barrier:

- WORK OUT AT HOME AT FIRST. If you really feel self-conscious about working out in public, find an exercise video or exercise show that you like, and follow along in the comfort of your own living room. This will help you feel more confident and competent and show you that you deserve to be outside on the paths or in the gyms just like everybody else.
- SCHEDULE AN APPOINTMENT WITH A TRAINER. Most gyms offer free or low-cost sessions with a trainer for those just starting out. A trainer can show you how the machines work and help you create a training plan that is personalized to your own conditions and goals. This will take away some of the mystery and build up your confidence.

If you continue to feel that you have an image problem, you may want to seek the help of a professional psychotherapist or psychologist specializing in body image. The government Web site healthfinder.gov has links to several Web sites that offer information and professional referrals on body image and eating disorders.

"It's Too Much; I Don't Know Where to Start"

It's easy to feel overwhelmed by these negative thoughts: "I have tried so many times in my life to lose weight, eat right, and exercise more, and I am not getting anywhere." Living with mindfulness means that you are not dwelling on your past. The past is already gone. You have the power to not repeat the past habits that did not work for you. You have the choice to follow a different path, a different way of living that is aligned with your intention. By now, chances are that you already have a good picture of why you failed with all your previous attempts. Take a few deep breaths, relax, and know that you don't have to do everything at once. A small change here and there builds up over time.

Consider this suggestion:

- MAKE PROGRESSIVE GOALS, AND KEEP TRACK OF YOUR PROGRESS. You need to start somewhere, so the best thing to do early in your journey is to set an achievable goal and then build from there. If your ultimate goal is to walk ten thousand steps per day, make your first goal three thousand steps per day. Keep track of your progress every day, recording how many steps you've taken in your exercise log, and once you have met that goal and maintained it for little while, increase your goal to six thousand steps per day and build up again. You can keep track of your goals and progress by using a notebook and pen, or one of the many online tools—whichever you like best. The important thing is being consistent with your activity and your tracking.

"Working Out Is Too Expensive"

Just as you don't need to run marathons to benefit from exercise, you don't need to spend thousands of dollars on clothes, equipment, and gym memberships to get the exercise you need. When it comes down to it, the only thing you really, really need is a good pair of walking or running shoes.

Consider these ways to work around this barrier:

- EXPLORE THE YMCA OR COMMUNITY CENTERS. Although you don't need a health-club membership to get the exercise you need, it can be nice to have access to workout equipment, classes, and trainers. Some clubs can be incredibly expensive, but there are a lot of them that aren't. YMCAs are generally very reasonably priced, as are community-center gyms, and often these facilities offer a sliding pay scale depending on how much people can afford.

- INVEST IN YOUR SHOES. You can spend thousands of dollars on exercise clothes and equipment, but really, you don't need them. Comfortable, sweat-wicking clothes can be nice, but these don't have to be expensive, and you don't need a different outfit for every day of the month. What you really can't skimp on are walking/running shoes. They support your feet and help keep your joints happy. Check with your network of friends or colleagues at work to figure out which ones will give you the best value for the support you need for your exercise routine.

"There's No Place to Exercise in My Neighborhood"

This is a problem we have all dealt with at some point in our lives, maybe for all of our lives: lack of sidewalks or open spaces; no health clubs close enough to home. It's hard to get a good workout if there's simply no place nearby to exercise. Yet there are ways to overcome this barrier.

Consider these suggestions:

- WORK OUT AT HOME. It may not be your first choice, but working out in your living room or bedroom can be just as good as doing it outside. It just takes a bit more planning and the right type of equipment—whether it's an exercise bike, treadmill, or good step-aerobics video. In fact, research has shown that having access to home exercise equipment can actually help boost the amount of exercise we get.[32]
- PLAN WORKOUTS AWAY. If there are good places to exercise away from your neighborhood—whether at work, school, or a friend's place—plan a little extra time and get in a walk or jog early in the day or before you head home.

- WORK WITH YOUR COMMUNITY TO GET TRAILS AND PATHS BUILT. This is a long-term process, not a quick fix to your problem of finding a place to be physically active. But we can all work to make our communities healthier. It can be as easy as sending an e-mail or talking to your local legislator, or as involved as organizing a presentation to the city council. Whatever you can do, it's a great move that will benefit not only you but also your loved ones and everyone else in the community.

"My Neighborhood's Just Not Safe"

The unfortunate reality is that we sometimes live in neighborhoods that just aren't that safe, and this can be a big barrier to getting the exercise we need. Not only does it cut down on our opportunities to be active; it also saps our energy as we try to figure out ways around it, all the while thinking about our safety and the safety of our loved ones.

Consider these ways to work around this barrier:

- FIND A SAFE INDOOR AREA TO EXERCISE. This can be your home, a community center, or a local YMCA. Just find which one, or which combination, works best for you. Videos or TV fitness shows can get you moving at home, and community centers, YMCAs, and other types of gyms can offer access to exercise equipment and regular exercise classes.
- EXERCISE WITH A GROUP. There actually is safety in numbers. Find some folks with goals similar to yours, and get outside for walks or jogs together a few days a week. Not only will you feel safer; you'll be more motivated to get out there, since your group will be depending on you. You can make a commitment to others to exercise regularly with them. Join a dance, aerobics, or martial-arts class. It may be easier to stay motivated and interested in our exercise if we do it in a group setting.

- GO WHERE YOU FEEL MORE COMFORTABLE. If you can, try to go to areas where you feel more comfortable to get your exercise. Maybe this is during lunch at school or work, using the stairways or the grounds nearby. Or you can make a special trip to somewhere beautiful just for a healthy change of pace.

"My Body Has Aches and Pains"

People who have chronic conditions such as osteoarthritis, high blood pressure, or type 2 diabetes may worry that physical activity would make their condition worse. The *2008 Physical Activity Guidelines for Americans* recommends regular physical activity to promote quality of life and reduce the risk of developing other new chronic conditions such as heart disease.[33] It is essential, however, that people with chronic conditions consult their health-care providers about the type and amounts of activity that they should routinely engage in, and get monitored by them regularly.

The Mindfulness Connection

Now that we have discussed the science behind the link between exercise and health, and some of the nuts and bolts of fitting exercise into your life, it's time to connect activity with the practice of mindfulness—something that will make integrating activity into your life that much easier and that much more meaningful.

The journey toward weight loss is a journey of mind and body, and in no place is this union more clear than in the act of exercise and moving—mindful moving. When we are active—whether we're walking, climbing a mountain, or working in a garden—we are bound to the moment, which is the essence of mindfulness. If we weren't really in the moment, we'd stumble, we'd lose our footing, or we'd dig up the wrong plant. And by being in the moment we connect more closely to our senses, to our meditative breathing, to our bodies in general.

All these connections ground us and help us transcend the daily stresses, the daily barrage of commercials and other hurtful stimuli, giving us the peace of mind and body that helps us lose the weight we want, achieve the healthy life we need, and touch peace.

There is no better demonstration of the power of mindfulness than the walking meditation. Walking meditation is one of the ways to contemplate peace. Walking generates the energy of peace, solidity, and freedom.

Walking Meditation

Take two or three steps for each in-breath and each out-breath.
Breathing in, say, "I have arrived"; breathing out, say, "I am home."
Breathing in, say, "In the here"; breathing out, say, "In the now."
Breathing in, say, "I am solid"; breathing out, say, "I am free."
Breathing in, say, "In the ultimate"; breathing out, say, "I dwell."

To start, when you breathe in slowly, take two or three steps. Bring your attention to the soles of your feet, and become aware of the contact between your feet and the ground. Bring all your attention down to the soles of your feet.

Breathing in, take two or three steps, and say with each step, "I have arrived."

Breathing out, take another two or three steps, remaining mindful of the contact between your feet and the ground, and say, with each step, "I'm home."

But, arrived where? And where is home? you may ask.

According to the teachings and practice of the Buddha, life is available only in the present moment, in the here and now. And when you go back to the present moment, you have a chance to touch life, to encounter life, to become fully alive and fully present. That is why every step brings us back to the present moment—so that we can touch the wonders of life that are available. Therefore, when you say, "I have arrived," you're saying, "I have arrived in the here and now—the only place, the only time when life is available, and this is my true home." "I have arrived" means, "I don't want to be distracted and lost anymore, because I know that conditions for my happiness are already here in the present moment." Sometimes we believe that happiness is not possible in the here and now, that we need a few more conditions to be happy. So we run toward the future to get the conditions we think are missing. But by doing so we sacrifice the present moment; we sacrifice true life.

In our daily life, we have a tendency to think about the past, to get caught in sorrow and regret concerning the past, and to get caught in fear and uncertainty about the future, so our mind is not in the present moment. That is why it is very important to learn how to go back to the present moment in order to become fully alive, fully present. Walking meditation helps us do that easily.

WE NEED TO LEARN to walk in such a way that every step brings us stability, freedom, healing, and transformation. In order for each step to be solid, to be free, to be healing, to be nourishing, we need the energy of mindfulness and concentration, which is obtained by mindful breathing and mindful walking. "*I have arrived. I am home.*" That is not a statement. That is a practice, and only you can know whether you have arrived in the here and now; no one else can tell you. If you are truly established in the here and now, you feel free, light, and peaceful, and you can get in touch with all the wonders of life that are available.

Walk in such a way that peace becomes a reality in every cell of your body, in every cell of your consciousness. When we breathe peacefully, the peace of our breath penetrates into our body and mind. Then very soon, in no time at all, body, mind, and breath become one in concentration, and we get the energy of stability, solidity, and freedom generated by every step we make.

After a few minutes, you may move to the third line of the meditation: "*In the here. In the now.*" It means I am at home in the here and now. The address of the kingdom of God, the address of peace and togetherness is the here and now, not in the past, not in the future, and not in any other location in space.

After some time, you might like to move to the fourth line: "*I am solid. I am free.*" Solidity and freedom are the most important aspects of happiness. Without some solidity, without some freedom, true happiness is not possible; therefore, every step should be able to generate more solidity and freedom. And again, this is not a wish or a declaration. It is a practice.

So, "*I am solid, I am free*" means I notice that now I am more solid, I am freer. That makes the practice much more pleasant. You walk with dignity, like a king or queen. You walk like a lion, because you are truly yourself, with all your serenity and strength.

Every step becomes a delight. Every step has the power to heal, to transform. Not only can we heal ourselves with our steps, but we can help heal our loved ones, friends, the earth, and the environment. As we walk with mindfulness, we become calmer. Our brain becomes clearer and more lucid, allowing solutions to emerge to whatever pressing challenges we ourselves, our loved ones, our friends, or our world face.

The last line of the meditation is "*In the ultimate I dwell.*" There are two dimensions to reality. The first is the historical dimension, and the second is the ultimate dimension. We have an ultimate dimension—the ground of our being—and if we know how to live deeply every moment of our historical dimension, we can touch our ultimate dimension.

The historical dimension can be likened to a wave. And the ultimate dimension is like the water in a wave. A wave may seem to have a beginning and an ending. A wave may be seen as high or low, big or small, different from or similar to other waves.

But the appearance of beginning and ending, high and low, more or less beautiful, belongs only to the historical dimension. Because the wave is at the same time the water. Water transcends the form of the wave, and the idea of beginning and ending, high or low, this wave or that wave. These notions apply to the wave but not to the water. The moment when the wave realizes that she is water, she loses all her fear and she enjoys much more being a wave. She is free from birth and death, being and nonbeing, high or low, because when we are able to touch our ultimate dimension, we are no longer subjected to fear—fear of being less than or more than other waves; fear of birth; fear of death.

This is a very deep practice. When you touch your true foundation, your true nature—the nature of no birth and no death—then nonfear arises. And with nonfear, true happiness becomes possible.

It is possible to live each moment of our daily life in a way that helps us touch our ultimate dimension. In fact, it is *only* through living our daily life in the historical dimension deeply that we can dwell in the ultimate.

Touching the Sky and the Earth

To develop concentration, physical stability, strength, and flexibility, Thay regularly practices Ten Mindful Movements as described in the book *Mindful Movements: Ten Exercises for Well-Being.* They are simple movements based in yoga and tai chi. One of them is called Touching the Sky and the Earth.

> Your feet are hip-width apart with knees slightly bent. Breathing in, bring your arms up above your head, palms forward. Stretch all the way up, and look up as if you are trying to touch the sky. Breathing out, bend at the waist as you bring your arms down to touch the earth, or as far as you can comfortably reach. If your hands can touch the ground, feel your hands planted into the earth. Release your neck. From this position, breathe in, and keep your back relaxed as you come all the way back up and touch the sky. Touch the earth and sky three more times.[34]

Mindful Stretching and Fidgeting

Fidgeting is a good way to burn off excess calories throughout our waking hours. Research has shown that we can burn calories outside formal physical activity regimens. Fidgeting—the little movements that we make such as pacing as we talk, gesturing with our hands, standing instead of sitting—can burn as much as a few hundred calories a day.[35] Instead of sitting down on a chair while you are talking on the phone or talking to someone, try pacing around instead. While waiting in line at the bank or grocery store, while waiting for the bus or sitting at a red light, you can mindfully stretch, pointing and flexing your feet or flexing and relaxing your legs and arms. Or you can mindfully fidget by jiggling your limbs as you stay with your conscious breathing.

Increasingly, we are spending more hours sitting in front of a computer screen. We communicate with others via e-mail and blogs; we search the Web, watch movies, work on our assignments; and so forth. Try setting your computer screen higher so that you can stand up while using the computer. This way, you can take regular breaks to stretch and use more of your muscles, burning more calories. As you stretch and move around, you can also prevent back pain and shoulder aches caused by sitting and typing for extended periods.

Breathing in, I fidget and pace around.

Breathing out, I prevent my body from stiffening/
tensing up.

Mindful Moving with Other Activities

While mindful walking is the perfect way to begin the practice of mindful moving, you certainly don't need to limit yourself to walking if you like other activities just as much, if not more. Whether it's cycling, dancing, running, gardening, or doing yoga, you can apply the same meditative practice used in mindful walking.

If you're riding your bike, every ten pedal strokes use a line from the meditation along with your in and out breath: "I have arrived. I am home."

If you're working in a garden, every other time you dig with your spade or pull a weed, breathe in and practice: "I am solid." Breathe out and practice: "I am free."

It may seem a bit forced and odd at first, since most of us have never approached activity with meditation in mind. But as you practice, it will become an integral part of your activity, not only helping you ground yourself in the present moment but making exercise an experience you'll enjoy like you never have before. It will become a focal point of your day, when you'll be able to connect with the moment and connect with yourself, and it is through this connection, and the dynamic nature of exercise and movement, that you will be able to lose the weight you want to lose and become the healthy person you want to be.

Mindful Arm Swing

A Movement to Help Untie Your External and Internal Knots

This movement comes from the traditional Chinese *chi qong* practice. It is both releasing and invigorating. It can be done wherever you are. It does not require any equipment; your own body is the instrument. It is convenient, since you can do it anywhere, at any time, and it takes only five minutes.

1. Stand with your feet at shoulder width; relax the body with the knees slightly bent.
2. Have your eyes focused on an object or scenery in front of you.
3. Swing your arms up straight in front of you toward the sky or ceiling, as you inhale deeply.

4. Drop and swing your arms back down all the way and behind you, as you exhale completely.
5. Repeat this up-and-down movement continually.
6. Increase the speed of the up-and-down movement gradually throughout the exercise for five minutes.

When doing this exercise, try to feel that you are "swimming" in air. You are one with the air, and you exchange energy with air. When you move the arms up and inhale, you are taking in fresh energy from all that is around you, and when you swing your arms down on the exhale, you expel all the burdensome energy. Every move is a mindful move, and every move is actively engaging with the air and your breath. You will instantly feel different after doing the arm swing for five minutes continually. Your heart is pumping, and you feel happier. The tensions in your head and around your shoulders and back muscles start to release and relax. The movement together with the breathing will take you back to yourself, uniting body and mind. If you have back problems or other physical concerns, please consult your health-care provider before starting this exercise.

Putting It All Together

We have covered a lot of ground in this chapter: the health and weight-loss benefits of exercise; the amount of exercise needed to lose weight and keep it off; the important connection between mindfulness and activity; and the common barriers to getting the exercise we need. Now it is time to put it all together and create a plan for our active selves. Let's call it your *in*Moving Plan—where *in* conveys being in the present moment. With your *in*Moving Plan, you will be able to set your goals, figure out ways around any issues that might keep you from reaching your goals, and plan the activities you want to do to reach your goals.

Your *in*Moving Plan

Here are the key elements to your successful plan for increasing the amount of activity in your daily life. Go through these elements here, and then make your own action plan, following the example at the end of this section.

Like all action plans, your goals are not meant to be a static document. As you gain experience and work through barriers, your goals and tips for overcoming barriers may change. Don't be afraid to adjust things as long as you make sure you stay committed to living mindfully, staying active, and working toward your ultimate goal of losing weight and keeping it off.

Why do you want to be more physically active?

Think about the reasons you want to be more active. They can cut across all parts of your life.

Examples: *I want to feel better about myself. I want to lose weight. I want to feel more energized.*

What's bad about being sedentary?

Think about the downsides of remaining inactive or not getting enough activity. Again, they can cut across all parts of your life.

Examples: *I'll stay heavy. I won't feel good about myself. I'll feel down and sad.*

What activities do you like to do?

Think about activities that give you joy. A lot of different things can count toward your exercise time. Be sure to choose activities that you love and have fun doing. If you feel as though you don't enjoy doing anything active right now, pick the least objectionable activity, or pick one for which you have a lot of social support, such as walking with a close friend.

Examples: *walking, biking, gardening, golfing, dancing, yoga, hiking, basketball, tennis, martial arts, bowling, skating, active play with your children or grandchildren, skiing, swimming*

What are your time goals for being active each day?

Ultimately, you want to get about thirty to eighty minutes of moderate activity (or up to about ten thousand steps) every day. Unless you're close to that right now, though, you don't want to start with that goal. It's best to build up to it, starting with one or two easier goals that you can use to step up to your overall goal. Maybe you want to start with twenty minutes a day (or twenty-five hundred steps) to start. Once you have done that for four weeks, you can move the goal to forty minutes a day (or five thousand steps). Then after another successful four weeks, you can work toward sixty minutes a day (or seventy-five hundred steps).

Example:

Goal 1: 2,500 steps/day (20 minutes/day)

Goal 2: 5,000 steps/day (40 minutes/day)

Goal 3: 7,500 steps/day (60 minutes/day)

What are two or three obstacles that can make it hard for you to be active? What are a couple ways around each of these?

Finding your way around obstacles that get in the way of being active is a necessary part of the process, and this is so for the Olympic marathoner as well as for the person just starting out. Think about some of the key things that keep you from getting out the door to get the exercise you need. They may be part of the list of barriers we went over earlier in this chapter, or they may be something else entirely. Once you have the top two or three, write them down and then come up with ways you can get around them. This will be your go-to list when things get tough and you are looking for reasons to not get out the door.

Example:

Obstacle: I don't have enough time.

My tips: I'll lay my clothes out the night before and get up early before the rest of the family.

I'll bring a sack lunch to work and go for a walk after I eat.

What are your mindful moving words?

Choose a line or two (or all the lines) from the walking meditation (see p. 170) to focus on while you are doing your exercise of choice.

Example: *"I have arrived. I am home."*

Mindful Moving: An Opportunity to Help Our Planet Stay Green

As we consciously use our body more and come to depend less on gadgets and automobiles for our chores and getting around, we not only burn more calories but also contribute to reducing our ecological footprint. There are many ways that you can move more in your daily life. Use the stairs instead of the elevators or escalators. Bike to work. Walk or bike to places within five miles. Use a push mower to cut your lawn and a rake to gather leaves. Hang some of your laundry out to dry.

In the book *The World We Have,* Thich Nhat Hanh talks about the practice of a No-Car Day once a week in his monasteries and practice centers as a way to reduce carbon emissions and gas consumption.[36] Practicing No-Car Days can give us much joy. We can each do something concretely on a regular basis to protect the planet and reduce global warming. Spread the joy and encourage your family, friends, and colleagues to pledge themselves to a No-Car Day once a month or once a week to start. (For more information, visit www.carfreedays.org.)

The Bottom Line

For many of us, exercise is really challenging. It's not easy to find the time. It can be unpleasant, especially early on. And there are a lot of competing activities that may be less taxing and seemingly more appealing. But our bodies and minds truly crave activity. For thousands and thousands of years, humans have been active beings, and this has been hardwired into our genes, our cells, our minds. The fact that today's modern life has stifled this side of us doesn't mean it has faded away. To truly become ourselves, to truly realize who we are, and to appreciate the physical capacity that we have been uniquely endowed with requires us to reconnect with our active selves, however out of touch with them we may be.

This doesn't mean you need to scale mountains or run marathons. It's much simpler than that. It simply means you need to spend some time most days of the week doing physical activities you like. It's a practice in mindful moving that will help you transcend the daily grind, connect with yourself, and reach your healthy weight while contributing to the well-being of our world.

CHAPTER 7

Mindful Living Plan

NOW THAT WE'VE TALKED about the roles that food and physical activity can play in our mindfulness practice and our journey toward a healthy weight, it's time to take a step back and take a broader view of how we can incorporate mindfulness into our lives.

Mindfulness practice touches the stillness in ourselves. It allows us to calm down and reflect so that we can reconnect with our true self. Our true self has been camouflaged by our numbed, autopilot way of life, our days overloaded with countless daily demands and by the never-ending stimuli from our high-tech, advertising-driven consumer society. When we are free from our automatic responses, we can see more clearly things as they are, from moment to moment, without judgment, preconceived notions, or bias. We get to know ourselves better. We become much more in tune with our own feelings, actions, and thoughts as well as with the feelings, actions, and thoughts of others. As we live in each moment fully, we learn to love ourselves and to make peace with all that is around us. In sum, we simply *savor* life.

Most people can't practice healthy living even though they know they should. There are innumerable inner and outer barriers we can name. To transcend these, you need to ask yourself what it is that you *really want.* Often our habit energy and fear prevent us from identifying what we want and from living healthily. Habit energy keeps us going, but we may not know where we are heading. We struggle even during our sleep. We need to practice mindfulness in our daily lives to check and transform the destructive habit energies that are taking our lives in the wrong direction.

With the creation of your healthy-weight mission statement, as described in chapter 1, you have already taken the first step on the mindful journey to improving your health. As you practice with the intention to improve your eating habits, you will find that the same mindful eating practices can change your perception of everything else that you do and experience. Similarly, if you learn to approach other aspects of your life with mindfulness, you will find that these good habits can in turn enhance your efforts to eat more healthily.

Practice focusing on the now and immersing yourself entirely in the task at hand, whatever it may be. You will find that you can complete the task with less effort. The way you engage in the task will completely determine the quality of the future. How the future unfolds depends on the way we handle each present moment. Being mindful from moment to moment gives us the greatest opportunity to create a successful and beautiful future.

Chapters 5 and 6 covered how to shine the light of mindfulness on the everyday activities of eating and moving, and guided you through the creation of the *in*Eating and *in*Moving strategies to help you chart your mindfulness course. To jump-start you in your mindfulness journey, we suggest a mindfulness practice plan to help put together some of the exercises and recommendations that we have talked about throughout the book and integrate them into your daily routines. The proposed plan includes practices not only in eating and moving for weight control, but also in transforming and enjoying life. We call it the Mindful Living Plan. It has three main components: *in*Eating, *in*Moving, and *in*Breathing. As we discussed, *in* denotes "in the moment." These *in* strategies can be personalized and integrated seamlessly into virtually every act of your daily living, becoming the pillars for helping you build a mindful life.

While *in*Eating and *in*Moving specifically address food and physical activity for health, the *in*Breathing strategy addresses all other aspects of what we do and helps us transform our habits and afflictions. The *in*Breathing strategy helps awaken all our senses and helps us be fully present to understand and skillfully handle our thoughts, feelings, words, and actions.

When we eat, we eat mindfully. When we exercise, we exercise mindfully. We also look, listen, talk, touch, feel, think, and perceive mindfully. We are breathing all the time, and being aware of our breathing is the easiest and the most effective practice to get us focused on the present moment. Breathing is a core complement to both eating and moving mindfully as well as to practicing mindfulness throughout the day.

When we live with awareness, we will gain insight and understanding, diminish ignorance, and bring about love, compassion, and joy. Understanding the interdependent and impermanent nature of all things is the key to transformation, and mindfulness energy is the source of power to fuel the transformation process in every moment.

Just as sunlight provides the energy for a seed to grow into a plant, mindfulness provides the energy to transform all mental formations—our mind states, which are expressions of seed energies manifested in our mind. Mindfulness energy is like the sun: it has only to radiate its energy to do its work naturally. The essential point is that we do not try to repress our afflictions, our negative energies, because the more we resist or fight them, the stronger they will grow in us. We need only learn to recognize them, embrace them, and bathe them in the energy of mindfulness. When there is an abundance of mindfulness energy in us, it can transmute and dilute the effects of the negative seed energies in us. Cultivating mindfulness energy will soothe and calm our negative emotions. The Mindful Living Plan provides a practical framework to build up such mindfulness energy.

When we feel stuck and immobilized, we are in our own way. Our ego is obstructing us. We are anxious, limited in our thinking, reactive, doubtful, and always thinking and worrying about the future or regretting the past rather than being with what's now. We become our worst enemy as we are overwhelmed by our fears, our anger, and our despair. These fears, anger, and despair are illusions. They are not real, but we believe they are real and let them dominate us. Take a few in-breaths and out-breaths so that your mind and your body are united again, allowing you to be fully in the present. Once you can be in the present, you will recognize that your fears, anger, and despair are all projections from the past. They are not the present reality.

This practice plan is only a guide to help you improve your well-being. These practices are not rigid formulas but are simply exercises to help you take the first steps to get a taste of mindfulness practice, to gain better insight, and to remove the clouds that cover our clear vision. As we learn and put these concepts into practice, it is important that we are not limited by them but learn to use them to gain greater understanding. Certainly, you can come up with your own practices by simply applying mindfulness principles in ways that are relevant and appropriate to your own life. The important thing is that you start taking the first few simple steps and see for yourself their effect. Continual daily practice will certainly build up your mindfulness energy. In the following sections we describe the three main components of the Mindful Living Plan: *in*Eating, *in*Moving, and *in*Breathing.

*In*Eating

In chapter 5, we discussed the key components of a mindful, nutritious eating plan, one that is good for you as well as our planet, paying attention to not only what you eat but how you eat. As you gradually increase your practice of mindful eating, you may find that you will be more in tune with your hunger and fullness cues. You may find that you will eat when you are hungry and stop when you are full. You may find that you will make nutritious, wise, and green food choices that are satisfying to you and good for our planet. You may find that you will eat with more awareness and deeply enjoy the choices you make.

At least once a day, eat one snack or one meal without any sensory stimulus other than focusing on the food and drink you are consuming. This means no television, no newspaper, no book, no radio, no iPod, no mobile phone, no thinking or worries. Eat slowly, really enjoy the food, and chew well to savor the taste of the food and help with digestion.

Take time to review the key nutrition principles in chapter 5 as well as the seven habits of mindful eaters, and check which ones you already have in place and which ones you need to change and improve. As we discussed in chapter 5, identify those that are most important to you so that you can come up with a list of food goals and mindful-habit goals. Focus on a new food goal and a new mindful eating practice every one or two weeks, and consciously adhere to them on a daily basis. Take note of the challenges and obstacles that prevent you from reaching your intended goals, and plan strategies to overcome them.

For example, if your food goal is to eliminate sugar-sweetened beverages, be aware of your habit energy leading you to reach out for a soda throughout the day. As you go toward a sugar-sweetened soda, stop yourself, pause, take one in-breath and out-breath, and say silently to yourself, "I made a commitment to eliminate sugar-sweetened soda. I am going to have lime-flavored sparkling water instead." If your mindful eating goal is to choose smaller portions, use smaller utensils to serve yourself, and use a smaller plate. As you adopt a new healthy eating habit and a new mindful eating habit every week or two, over the course of four, eight, or twelve weeks you will find your daily eating practice to be much more healthful and in line with your healthy-weight goals. Since you are the one who decides which healthier choice you will incorporate into your own daily eating plan, the likelihood of you sticking with your choice is much greater than if the incentive for change were to come from someone else.

*In*Moving

Even if we may no longer work in the fields like our ancestors, there are many opportunities for us to move our bodies every day. As we discussed in chapter 6, physical activity is one of the best means of practicing mindfulness, because it is so connected with body, mind, and the here and now. Our bodies (and our minds) yearn to be active. The key is finding routines that you love and that can be integrated into your everyday living.

At least every day, find a route or hallway you can walk mindfully. It can be in your home, at work, in a park, or somewhere around town where you walk often. As you walk, put your full attention on your feet and their contact with the ground. You walk for the sake of walking, not to reach any destination. You walk without any "to do" list, regrets of the past, or worries about the future. You breathe as you walk. For a slow pace, take two steps for each in-breath and out-breath. For a brisker pace when you need to reach your appointment on time, take three steps or more for each in-breath and out-breath.

To gain the physical, emotional, and weight-loss benefits of increased activity, review your *in*Moving plan from chapter 6. What goals did you set for mindfully increasing your activity, and what strategies did you develop for overcoming your barriers to being more active? Use the goals you set to fill in the *in*Moving section of your Mindful Living Plan.

As you build up your practice of mindful moving, you will find yourself wanting to move more and more. And mindful moving will become a habit—a good habit. When you skip your mindful movement, you will find yourself missing it, even just for one day.

*In*Breathing

*In*Breathing involves every aspect of our lives. We have discussed the importance of conscious breathing and mindfulness practice, and we have given many exercises in chapter 4 that are wonderful practices for the *in*Breathing strategy. We can only describe the principle in the practice; it would be impossible to list all the different ways to practice. However, to help you instill more mindfulness into your daily actions as well as sustain mindfulness as a way of life, we suggest that you consider trying some of the exercises in chapter 4, and the additional meditations and verses in this chapter. These are what we call "breathing meditations in action": you can use them as you engage in various tasks and routines throughout the day. Pick and choose those that are relevant and appealing to you.

One of the aspects of engaged mindfulness practice is that it is *continuous.* We don't just have periods of mindfulness throughout the day: we want to be mindful all day long, as much as possible. When we incorporate mindfulness from moment to moment, we stay fresh, peaceful, and protected from the push and pull of our habit energy. We stay on track because we are awake, no longer on autopilot. The meditations and verses given here are like signposts along the road, reminding us of the speed limit, giving us directions, helping us stay on track. And we need them all throughout our journey, not just at the beginning or end. So it is with mindfulness. We may start out our day very mindfully, but if when we get in the car or bus to go to work we allow stress or worry to overwhelm us, we have lost our way. So we need practices throughout the day to help us remember to come back to our breathing, relax, and stay present in this moment. In this way, we will not be the victim of our stress and worry.

As you practice these meditations and verses, keep in mind that the very first step in opening your heart to mindfulness is to *stop* for a brief moment as soon as you are aware of your action. This brief stop invites you to look deeply and gain insight into what you are doing by being totally present at that moment. It is possible to turn every activity we do into a meditation, even the most mundane things—like using the toilet, brushing our hair, or putting on our clothes—as long as we have awareness, with our mind and body united.

Waking-Up Meditation

As you wake up every morning and before you get out of bed, breathe with the waking-up meditation. Breathe three in-breaths and out-breaths, repeating the following verse silently for each in-breath and out-breath.

> Breathing in, I fill my new day with
> joy/faith/love/gratitude/mindfulness/ease/harmony.
> Breathing out, I smile.

For the in-breath, choose one of the words that most appeals to you.

Sunrise Meditation

Sunrise is a very special time of the day. It is transient, brief in duration, yet magical if we can get up early enough to see it. It is the beginning of a brand-new day. Dawn is a reminder that we can start our life anew, unburdened by the troubles and worries of yesterday. The energy from the emerging rays nurtures all there is on Mother Earth, including plants, animals, and ourselves.

Keep your window clear, and watch the sky become gradually lit up with the rays of dawn. Witness the beauty of sunrise. As you see the sun rising, breathe in and out a few times.

> Breathing in, I am aware of the sun.
> Breathing out, I thank the universe for the sun's energy and
> brilliance.

Teeth-Brushing Meditation

Since we brush our teeth a number of times a day, it is a great opportunity for us to practice mindfulness. Remember the advice from your dentist: brushing our teeth properly will foster gum health and the integrity of our teeth throughout our lives. Without healthy teeth and gums, we would not be able to chew well and enjoy our daily food.

As you brush your teeth, breathe in and out a few times. Do not think about your next assignment or errand. When you brush your teeth, focus just on your teeth and gums, nothing else.

> Breathing in, I am aware of my teeth and gums.
>
> Breathing out, I look after my teeth and gums.

Hurrying Meditation

It is inevitable that we may find ourselves in a hurry at times. As you cultivate mindfulness, though, and plan your day better, your tendency to hurry may very well decrease. Nevertheless, when you do need to hurry, it doesn't mean you need to forget mindfulness as you rush along. It is better to be mindful than not mindful as you hurry, if only to avoid accidents and mistakes. Mindful hurrying means that you know you are hurrying. In a sense, you embrace that you are hurrying. You focus on the task at hand and do it in a faster and more efficient mode. Mindfulness doesn't necessarily mean going slow. You can be fast and still be mindful—totally aware and relaxed. For example, if you are going from one building to another in a hurry, you want to make sure that you are using the shortest route. As you hurry, pay attention to your in-breath and out-breath.

> Breathing in, I am moving quickly.
>
> Breathing out, I am in the flow.

Smiling Meditation

A smile is the universal language for happiness. When we see the "smiley" face on stickers, shopping bags, or T-shirts, we spontaneously smile even though we may not be in the best mood.

It is important not to forget our own smile and the power it has. Our smile can bring much joy and relaxation to us and to others around us at the same time. When we smile, the muscles around our mouth are stretched and relaxed, just like doing yoga. Smiling is mouth yoga. We release the tension from our face as we smile. Others who run into us notice it, even strangers, and are likely to smile back. It is a wonderful chain reaction that we can initiate, touching the joy in anyone we encounter. Smiling is an ambassador of goodwill. As you smile, take a few in-breaths and out-breaths.

Breathing in, I smile.
Breathing out, I relax and touch joy.

Light-Switch Meditation

We turn on light switches many times a day, at home or at our office. Each time you turn on a light switch, pause for a moment and practice a few breathing meditations.

Breathing in, I illuminate this room with light.
Breathing out, I thank the electricity that is available to us.

As you leave the room, turn off the light to save energy. Pause again after turning off the light switch and breathe in and out a few times.

Breathing in, I am leaving this room.
Breathing out, I am mindful of not wasting any electricity.

Sky Meditation

Staring into the sky, we see and feel the expansiveness of space, the ever-changing scenery that captures our imagination, the power and vastness of nature, and the smallness of ourselves in relation to our universe. Staring into the sky offers us a wonderful opportunity to free ourselves from the burdens of our daily demands and from our ego. It allows us to appreciate the reality of constant change and to be free to dream.

> Breathing in, I see the magnificent sky.
> Breathing out, I feel free.

Jogging/Brisk-Walking Meditation

Jogging or brisk walking is a great form of exercise for our cardiovascular health and weight control. Jogging unmindfully not only deprives you of the joy of jogging but may even be detrimental, because it can lead to injuries or accidents. As you jog or walk briskly, focus on your legs, feet, and what is in front of your eyes. If you are moving very briskly, you may shorten this meditation so that on the in-breath you say silently to yourself the word "moving," and on the out-breath you say to yourself the word "thanks." (However many steps you take on your in-breath, say "moving" with each step. And then however many steps on your out-breath, say "thanks" with each step.)

> Breathing in, I keep moving.
> Breathing out, I thank my body for its strength, endurance.
> and coordination.

Driving Meditation

Driving meditation allows us to focus solely on our driving, with no distraction from conversations with other passengers, no thinking, no talking on the cell phone, and no text messaging. Discussions with fellow passengers or someone on the phone can, with all good intentions, end up in involved or heated conversations that distract us from paying attention to what is happening on the road—and what is happening in the moment.

> Breathing in, I am driving my car.
> Breathing out, I am mindful of all that is around me.

Traffic-Jam Meditation

Many people become impatient and irritable when they are in a traffic jam. What can you do to speed up the traffic? Nothing. The traffic will have to take its own course. And this actually makes it a great time to practice mindfulness. It provides valuable time for us to decompress and go back to the island of calm in ourselves to get refreshed.

> Breathing in, I follow my in-breath.
> Breathing out, I follow my out-breath.
>
> Breathing in, I know everyone is trying to get somewhere.
> Breathing out, I wish everyone a peaceful, safe journey.
>
> Breathing in, I go back to the island of calm in myself.
> Breathing out, I feel refreshed.

Water-Faucet Meditation

In developed countries, clean sanitary water is available continuously from the water faucet. Yet 1.1 billion people worldwide are still without clean water.[1] As we turn on our faucet, it is a reminder of how blessed we are. An average American uses four hundred liters of water a day.[2] It takes twenty-four hundred liters of water to produce a hamburger, but only twenty-four liters to grow a pound of grain. We need to use water carefully and find ways to help others around the world have access to clean water—an essential for daily living.

Breathing in, as I turn on the faucet, I am grateful for the clean water that sustains my life.
Breathing out, I remember the billions of people who are without clean water every day.

Elevator Meditation

While we are waiting for an elevator to arrive, it is easy for us to become impatient and get irritated about the wait. This window of time is actually a great opportunity to sneak in some conscious breathing to help us maintain our calm and return to the present moment.

For those of us who are claustrophobic or afraid of heights, breathing in and out mindfully is a wonderful way to take care of our anxiety when it arises.

Breathing in, I am aware of my in-breath.
Breathing out, I am aware of my out-breath.

Breathing in, I embrace my discomfort.
Breathing out, I feel a lot of space and security inside me.

Greeting-Our-Negative-Emotions Meditation

It is human for all of us to have negative emotions on a daily basis unless we are a seasoned mindfulness practitioner who knows how to prevent and transform them. Whenever a negative emotion arises, be it anger, despair, sadness, frustration, fear, or anxiety, repeat the following *gatha* (verse) silently to yourself for three to six in-breaths and out-breaths.

Breathing in, I am aware of my anger/despair/sadness/frustration/fear/anxiety.
Breathing out, I embrace my anger/despair/sadness/frustration/fear/anxiety.

Name the emotion that is the strongest at that moment. The more often you can catch your negative emotions as they arise, breathe into them, and embrace them, the easier it is to transform them. What you are essentially doing is preventing your body from engaging neural pathways that produce stress hormones, which are good for you when you need to jump out of the way of an oncoming train but not helpful in the course of everyday life. The more frequently we can disengage the coupling of perceived stress, be it physical or emotional, to our body's stress response, the greater the likelihood that we can maintain our well-being. The route to transformation is really as easy as going back to our in-breath and out-breath—the action that we are all constantly engaged in as long as we are alive. The only task for us is to be aware of our emotions and reconnect to them through our in-breath and out-breath.

Internet/E-mail Meditation

The Internet and e-mail are now a way of life and a principal means of communication in the twenty-first century. It is quite easy to be totally consumed by the Internet, chained to our chair, forgetting to get up and move around, eat or be in touch with our body. After hours of staring at our computer screen, our eyes are strained, our back may hurt, our shoulders are stiff, and our mind can become numb.

We can refresh ourselves easily by breathing with awareness. The following meditation is very helpful in preventing major mistakes or disasters that occur when we are on computer overload, like sending sensitive e-mails to unintended recipients. You can use it whenever you write an e-mail before you click "Send."

Breathing in, I thank the power of the Internet.
Breathing out, I am fully conscious of my current e-mail actions.

Deep-Listening-and-Loving-Speech Meditation

Many of us find it difficult to communicate with our family members or colleagues. At times we can become intolerant of them and irritated with their views and advice. We lose our capacity to listen deeply to them or the willingness to understand their point of view. We cannot speak calmly with others, or when we do talk, our own suffering, fear, or anxiety surfaces, and our words become critical and bitter.

We need to learn the art of listening and speaking. To help restore communication, we need deep, compassionate listening to help us understand others better. This means that our only intention while listening is to help the other person suffer less and express what she has in her heart. We become completely present to just receive what she needs to share, without judging or reacting. Even if the other person says things that are not true, that contain a lot of blame and bitterness, we do not correct her straightaway. We give her space to share her feelings, and later on, maybe a day or two later, we can slowly share information that will help her release her wrong perceptions about us or the situation.

We also need to apply the methods of loving speech, using only words that inspire confidence, joy, and hope in others. We take time to acknowledge the positive, beautiful things in others. We let them know how important they are to us, and we thank them for the way they contribute to our life. We also can very mindfully and patiently express our difficulties in our relationship with them without judgment or blame. We take responsibility for our own feelings and reactions but ask them to support us and help us by watering the good seeds in us, not the negative ones. And we can ask them how we can be of support to them when they go through difficulties. This way, we can attain peace and harmony in our interactions.

Start with your family or those you are closest to. Before starting a conversation with a loved one, take a moment to breathe in and out a few times.

Breathing in, I listen deeply.
Breathing out, I speak with love.

As you form the habit of being able to listen deeply and speak in a positive, constructive manner with your loved ones, it will spread to other interactions you have with friends and colleagues.

Tree Meditation for Stability

A tree is an inspiring image of sturdiness. When there is a storm or very strong wind, we see the branches shaking and bending. Yet the trunk of the tree remains sturdy and still, with its roots firmly in the soil. When you are in a state of turmoil or feeling vulnerable, look for a tree. If there are none nearby, look at an image of a tree. You may want to hang a picture of your favorite tree in your room or office as a reminder to breathe in and out with the tree whenever you feel unsteady.

Breathing in, I am like the trunk of the tree.
Breathing out, I can maintain my sturdiness despite stormy circumstances.

Flower Meditation

When we see a blooming flower, its natural beauty and fragrance never fail to lift our spirit no matter which corner of the globe we are in. Flowers bring us much joy and are a universal expression of love and appreciation for others, both in celebrations and memorials. Yet a flower withers shortly after it blooms—a profound reminder of the impermanence of all life. Being with a flower mindfully is a deep meditation.

Breathing in, I am grateful for the beauty and fragrance of this flower.
Breathing out, I treasure the flower here and now.

Standing-in-Line Meditation

In our daily lives, we often find ourselves having to wait in line. Sometimes we get irritated by the wait. It can be the checkout line at the supermarket, the security-check line at the airport, or the pickup line at our child's school. Being in a line is a great opportunity to sneak in mindful breathing and refresh yourself.

> Breathing in, I use this time just for myself, to unite my body and mind.
> Breathing out, I feel refreshed.

Multitasking Meditation

Juggling many tasks has become a way of life for many of us today. It is particularly true for people who have children or older parents to take care of, or for people who have to work more than one job to make ends meet. Think about your daily "to do" list: going shopping, preparing for a meeting at work, going to the post office, making a doctor's appointment, writing thank-you notes, and so forth. As we increase our practice of mindfulness, we may become more conscious of what is realistic for us to accomplish in one day.

> Breathing in, I am aware that I am juggling numerous tasks.
> Breathing out, I am mindful that I can accomplish only so much in one day.

What actually helps us to be more effective when we have many things to do is to engage each task with our full awareness, not worrying about the next task that needs to be done. While we run an errand, we simply do that task with our whole being. Then when we return home, we tackle the next task with the same focus and concentration, not thinking about other tasks. This way, our mind stays relaxed and fresh, and we have more energy to accomplish the items on our list as well as greater flexibility and acceptance when we need to adjust our schedule or our "to do" list.

Key Meditation

We may often find ourselves searching for our keys, whether they are to our car, home, or office. It helps to have a designated key hook in our home. Nevertheless, we still search for our keys everywhere—in the pockets of our coat or pants, in our backpack or briefcase. After you use a key, pause and breathe in and out to remind yourself of where you are placing your keys.

> Breathing in, I am aware of my keys.
> Breathing out, I place my keys here.

Cooking Meditation

With our hectic, demanding lives, fast food, cafeteria meals, and take-out from grocery stores or restaurants can become the mainstay of our eating. Many of us no longer really cook. In New York City, with meal-delivery service everywhere, even breakfast is commonly delivered.

Cooking can offer a sacred time for relaxing our mind and nurturing our soul. The act of cooking inevitably involves consciously thinking about what you would like to eat, purchasing the right ingredients, preparing the food, and enjoying what you created. Before you start preparing a meal in your kitchen, breathe in and breathe out a few times to be in touch with the joy of cooking.

> Breathing in, I thank the universe for the wonderful foods available for this meal.
> Breathing out, I prepare this meal with love and joy.

Sunset Meditation

Every day the sun rises and the sun sets. Even if we do not have a full view of the sunset, seeing the light rays through a window around that time offers another precious opening for conscious breathing and renewal at the end of the day. As we enjoy the sunset and begin to wind down, we can reflect on our day and let go of the events of the day.

> Breathing in, I thank the sun for all its energy, which sustains all beings on earth.
> Breathing out, I will also do my part to support all of life and help reverse global warming.

Good-Night Meditation

Before you sleep, unwind with the good-night meditation with three in-breaths and out-breaths, repeating the following verse silently for each in-breath and out-breath.

> Breathing in, I release my worries/thoughts.
> Breathing out, I touch peace.

Ten-Week Sample Mindful Living Plan

Now that we have described the three pillars of daily mindfulness practice, it is time for you to put your *in*Eating, *in*Moving, and *in*Breathing strategies into a practical plan. On a daily basis, this means that you increase the practice of mindful breathing integrated into various daily routines. You have a healthy-eating routine that includes reducing the amount of calories you eat each day; you have a healthy-moving routine that includes increasing the amount of calories you burn in physical activity. And, you have a strategy to practice consolidating good habits as well as the mindful transformation of negative emotions. Set specific goals, and stick with them. If you find yourself falling off track, reflect on how and why you were not able to achieve what you intended. Through mindful awareness, reflect on ways that can support you in getting back on the right path. It will take time before these practices become second nature. However, the more you are aware, the more attention you pay from moment to moment in your daily living, the more you will find yourself increasing mindful practices throughout the day without consciously trying. Over time, the practices will become effortless, fulfilling routines that you love.

The ten-week sample plan will help you integrate the practical tools of mindful living into your daily life, with emphasis on weight control. The ten-week duration is only an illustration. You may need a shorter or longer time to practice, and you may choose the exercises you like to do. Throughout all the exercises, maintain conscious breathing as the foundation of your movement. For the *in*Breathing strategy, don't feel compelled to do all the exercises suggested during each two-week segment. But do keep adding more exercises from one interval to the next. The progression suggested in the ten-week plan starts with building up positive energies and progresses to recognizing and transforming negative habit energies so that you can remain steady on your path to good health. There is no rigid formula for success. The key to improvement is continuing to practice every day. (See table 7.1.)

Stay on Track

Staying on track means doing what you intend to do and staying committed. Do not underestimate your resilience and resourcefulness. You have the power and the ability. Your journey to a healthier weight is not a journey that you start and then give up. It is a journey that you are living every day for the rest of your life. However, it is natural for you to encounter obstacles throughout this journey.

Sometimes you may feel discouraged, thinking that weight loss is an unattainable goal. Whenever you encounter such negative feelings, breathe in and out a few times to come back to the present reality. You can then embrace your doubts with compassion and awareness and see past the illusion they create. You have the capacity to change and to overcome every obstacle or challenge you face on your healthy-weight journey. You may need others to support you, including your family, friends, coworkers, and doctor. Think of how they can help you to eat healthier and stay active. Then, be proactive and ask for their help.

We offer you a few other practical suggestions to help you stay on track. These include keeping a daily mindful living log, getting social support, having a good sleep routine, and doing something every day to nurture yourself and connect to nature.

Keep a Mindful Living Log

Research has shown that monitoring and tracking one's weight can help people lose weight as well as maintain their weight,[3] so weigh yourself every morning or once a week to know your weight. Recording one's eating and physical activity has also been reported to be strongly associated with weight control.[4]

Journaling is useful for exploring what you are ready for and what you are not. Journaling helps us look deeply. Find the most ideal time of the day for pondering your mindful living log. For some, it is the morning, when the mind is clear. For others, it may be the evening before going to bed. What's important is to find a window of time every day during which you can focus on your own body, your mind, your feelings, and your inward experiences so that you know what is going on with yourself. Use your daily mindful living log to track your progress and whether you are reaching your daily *in*Eating, *in*Moving, and *in*Breathing goals. Record your weight in the mindful living log as well.

You can also do this yourself online. Many Web sites offer online food diaries that automatically calculate your calorie intake. A free, and ad-free, site worth trying is MyPyramid Tracker (www.mypyramidtracker.gov), set up by the U.S. Department of Agriculture to support its new food pyramid. The site also has an exercise tracker.

Attackpoint is also a good, free Web site for tracking activity (www.attackpoint.org). It's geared toward serious athletes; but don't be intimidated: it's got a lot to offer everyone. Weight can be tracked as well.

Table 7.1 TEN-WEEK SAMPLE MINDFUL LIVING PLAN

Weeks	*in*Eating	*in*Moving	*in*Breathing
1 & 2	No more than two sodas per week; replace with water Establish a regular meal schedule and do not skip meals	Mindful moving—2,500 steps per day (or 20 minutes) or equivalent activities of your choice	Conscious breathing (p. 70) Smiling meditation (p. 190) Waking-up meditation (p. 188)
3 & 4	Eliminate white bread or rice; replace with 100 percent whole-wheat bread and brown rice or other whole grains Turn off the television and the radio during meals	Mindful moving—5,000 steps per day (or 40 minutes) or equivalent activities of your choice	Teeth-brushing meditation (p. 189) E-mail meditation (p. 194) Deep-listening-and-loving-speech meditation (p. 195)
5 & 6	Eat fresh vegetables or fruit at every meal Use smaller plates to encourage smaller portions	Mindful moving—7,500 steps per day (or 60 minutes) or equivalent activities of your choice	Cooking meditation (p. 198) Calming-the-body meditation (p. 72) Hurrying meditation (p. 189)

Weeks	*in*Eating	*in*Moving	*in*Breathing
7 & 8	Choose vegetarian proteins instead of red meat and processed meat Chew well and eat more slowly so you can savor your food	Mindful moving—9,000 steps per day (or 75 minutes) or equivalent activities of your choice	Standing-in-line meditation (p. 197) Water-faucet meditation (p. 192) Silent-meal meditation (p. 125) Greeting-our-negative-emotions meditation (p. 193) Good-night meditation (p. 199)
9 & 10	Choose olive oil more often than butter Tune in to your satiety—stop eating when you are satisfied, not overfull	Mindful moving—10,000 steps per day (or 80 minutes) or equivalent activities of your choice	Love meditation (p. 85) Embracing-a-habit-energy-with-mindfulness meditation (p. 15) Key meditation (p. 198) Traffic-jam meditation (p. 192) Light-switch meditation (p. 190)

Get Social Support

Getting support is important to your weight loss and weight maintenance. Recognizing the popularity of various diets differing in suggested amounts of fats and protein—like Atkins, Zone, Mediterranean, and South Beach—scientists conducted a sophisticated trial to ascertain different diets' effectiveness in inducing weight loss. Researchers were surprised that they did not find much difference among the various dietary approaches.[5] However, they found that when one is able to stick with the eating routine, it made a difference. Those who attended the group sessions lost more weight than those who skipped the sessions. These group sessions were designed to help keep participants motivated and informed while addressing their concerns. This finding suggests that behavioral, psychological, and social factors are probably far more important for weight loss than the relative amount of nutrients in a diet.

Other research has shown that social support can help individuals lose weight and sustain behavior change for weight control. Researchers found that weight loss can be greater when the spouse is included in the weight-loss program.[6] Others have found that couples can lose weight together even if only one of them is enrolled in a formal program.[7] Even more interesting, some researchers have observed that obesity appears to spread through social networks, such that our friends—of the same sex—may influence our tendency to lose or gain weight even more than our spouses. And our close friends may not be the only ones who influence us most with regard to weight gain and loss; more-distant acquaintances, the friends of friends, may also have some influence on our behavior.[8] The lead researcher on this latter study, Dr. Nicholas Christakis, suggests that these social interconnections can empower us to do good—both for ourselves and for others. "Even as we are being influenced by others, we can influence others. . . . And therefore the importance of taking actions that are beneficial to others is heightened. So this network thing can cut both ways, subverting our ability to have free will, but increasing, if you will, the importance of us having free will."[9]

Social support can come from your family, your friends, your colleagues at work, or even virtual online communities. Reflect and identify all possible support that you may be able to get among all the people you know in your social network—people who may also want to live more mindfully or are struggling with their weight like yourself. The support system does not need to be face-to-face. You may find support on a social-networking site. You may also want to start your own mindful living group with your own family, friends, or neighbors. Begin with even one person who can get together with you regularly every week. The group can grow over time as friends tell other friends. The important thing is that you have your network of people who share the same vision and are committed to the practice of mindful living.

Buddhist teachings greatly emphasize the role of the *sangha*. A *sangha* is a spiritual community of like-minded practitioners. A very powerful, collective energy of mindfulness is generated when people practice together, and this energy can support us and help us continue our practice when our own willpower is weak. This is why monks and nuns, as well as laypeople, in many spiritual traditions live together. And the Buddha often said that it is only in a *sangha* that you can realize the dharma, the teachings of the Buddha. This is why the Three Jewels—Buddha, dharma, and *sangha*—interact: when you touch one, you touch the other two.

Whenever people leave one of our retreats, we always encourage them to join a *sangha* in their local area or start one if one doesn't yet exist. This is the best way to ensure an energetic and joyful continuation of your practice after the retreat. Alone, we will quickly succumb to our usual habits and lose our mindfulness practice. *Sangha*-building is the noblest task, the most important thing a practitioner can do. And in the Buddhist tradition, we speak of Maitreya, the future Buddha. In fact, it may be that the future Buddha will become manifest as a *sangha*, not as an individual, because this is what the world needs. Individual awakening is no longer enough. A collective awakening is necessary for us to survive as a species.

If you are interested in connecting with others who practice mindfulness, there are hundreds of local *sanghas* all over the world. At www.iamhome.org you can find a worldwide list of *sanghas* in the Thich Nhat Hanh tradition.

Get a Good Night's Sleep

We mentioned in chapter 1 that research suggests that a good night's sleep may also be essential to controlling your weight. As a result of not having enough sleep, you may be too tired to exercise, or you eat more because you get hungrier when you stay up late. When you're tired you have trouble knowing what your body really needs, and when you don't sleep enough you simply have more hours in the day available to eat. Help yourself establish good sleep habits by going to bed at a regular hour to minimize fluctuations in your circadian rhythm, which is closely related to the release of hormones affecting your sleep. Avoid caffeinated drinks after midafternoon unless you want to deliberately stay awake late at night. Avoid eating a big, heavy meal late at night, since it can cause indigestion and disrupt your ability to sleep. Do not drink alcohol with the idea that it can help you sleep better. Though alcohol may act like a sedative or tranquillizer because it is a central nervous system depressant, it actually interferes with sleep and may result in insomnia.[10] Before sleeping, refrain from any stimulating activity including exercising, watching violent movies, or listening to loud music. Help yourself fall asleep peacefully by reading for pleasure, listening to soft music, and practicing your good-night meditation.

Do Something Every Day to Nurture Yourself

What are you doing to nurture yourself every day? Don't forget to love and care for yourself while you are loving and taking care of others. Loving yourself is the basis for loving other people. Be your own best friend.

Reflect on whether you have been really nurturing yourself, feeding yourself with the good nutriments for your physical body as well as your spirit. Consider not only what you have been doing to take care of your family members or your friends, but how you have been caring for yourself and for your own well-being too.

Do you know your passion? Are you doing the work you love? Do you know what you really want to do in this life? Are you doing what you want to do? Or are you losing yourself because you are trying to meet someone else's expectations? You don't have to be trapped in any predicament. You have the freedom and ability to lead your life the way you want.

Start nurturing yourself by identifying an activity that will help you refuel your enthusiasm and life force daily. Do things that will capture your spirit and bring you joy, watering your positive seeds in your consciousness. It can be something very simple, such as listening to your favorite song, appreciating a favorite drawing, watching the birds feeding from the bird feeder, staring at a beam of light from the sky touching the horizon, enjoying a beautiful flower, or chatting with a friend who has a great sense of humor. Don't just sit there and wait for your negative feelings to pass. Complaining will not change your life. Change your thinking, and you can let go of limitations you imposed on yourself. Explore, and be proactive.

Get Back to Nature

Writers from Lao-tzu to Ralph Waldo Emerson have urged humans to be in tune and in touch with nature. Many of us may have experienced that whenever we spend time with nature: whether we're in the forest staring at the trees or watching a peaceful pond, we are refreshed and feel better.

Research has shown that nature actually can have an impact on our mental functioning. According to environmental psychologists, urban environments can dull our thinking and affect our memory. If you walk down a street in the city, there are many signs and cues to draw your attention, like the threat of wayward cars or the attractive window displays. Paying attention like this takes energy and effort, according to these researchers. Our brain can get "directed-attention fatigue," resulting in greater distractibility and irritability. In contrast, as we are walking along a pond with trees, the images in nature enter our mind without provoking lots of mental activity, draining us of energy or triggering a negative emotional response. Our brain can relax deeply and replenish itself. This is why scientists think that immersing ourselves in nature—or even just viewing a tree or a patch of green from a window—can have a restorative effect.[11]

Prevent Relapses

It is normal for all of us to slip and go back to our old habits despite our best intentions. This can be very discouraging. How do we prevent relapse? This is a scientific topic that has been studied extensively in the treatment of addictive behaviors such as alcoholism.[12] Researchers identified three "high-risk situations" that were linked to nearly 75 percent of relapses they studied.[13] These three situations are negative emotional states, interpersonal conflict, and social pressure. The researchers proposed three strategies to cope with these high-risk situations—namely, coping-skills training, cognitive therapy, and lifestyle modification. In the domain of lifestyle modification, researchers recommended the use of meditation, exercise, and spiritual practices as ways to help support a person's overall coping strategy.[14]

When we practice mindfulness, we can become more aware and catch the moment when we first begin to slip. Slipping is just slipping. It doesn't mean we're sliding back to where we started, and it doesn't imply that we have failed. It is just a signal that we have to get back on the right course. Some routines may not be working quite right for you, so you'll need to tweak them to make them better. Or you may have been distracted by something happening in your life without being conscious of it. Every mindful moment is a new opportunity to prevent relapse. We can begin anew every moment. That's one of the real powers of mindfulness.

The Five Mindfulness Trainings are a good tool to prevent relapse and stay on course. (See figure 7.1.) They are also a concrete way to practice compassion, understanding, and love. They are a moral compass guiding us toward a healthy, happy life.

The first mindfulness training strengthens our innate reverence for life and reminds us of the suffering caused by the destruction of life, be it of the lives of people, animals, plants, or minerals. The second mindfulness training encourages our true happiness by reminding us that happiness can only be found in the present moment, through the practice of generosity and interconnectedness. It also raises our awareness of the suffering caused by exploitation, social injustice, stealing, and oppression. The third mindfulness training nurtures our capacity for true love and raises our awareness of the suffering caused by unmindful sexuality. The fourth mindfulness training invites us to care for ourselves and others through loving speech and deep listening and addresses the suffering caused by unmindful speech and the inability to listen to others. The fifth mindfulness training promotes our nourishment and healing through mindful consumption and is integral to our aspiration to attain a healthy weight. Recite the Five Mindfulness Trainings regularly—say, once a week—to support your mindful living.

Figure 7.1 THE FIVE MINDFULNESS TRAININGS

The Five Mindfulness Trainings represent the Buddhist vision for a global spirituality and ethic. They are a concrete expression of the Buddha's teachings on the Four Noble Truths and the Noble Eightfold Path, the path of right understanding and true love, leading to healing, transformation, and happiness for ourselves and for the world. To practice the Five Mindfulness Trainings is to cultivate the insight of interbeing, or Right View, which can remove all discrimination, intolerance, anger, fear, and despair. If we live according to the Five Mindfulness Trainings, we are already on the path of a bodhisattva. Knowing we are on that path, we are not lost in confusion about our life in the present or in fears about the future.

Reverence for Life

Aware of the suffering caused by the destruction of life, I am committed to cultivating the insight of interbeing and compassion and to learning ways to protect the lives of people, animals, plants, and minerals. I am determined not to kill, not to let others kill, and not to support any act of killing in the world, in my thinking, or in my way of life. Seeing that harmful actions arise from anger, fear, greed, and intolerance, which in turn come from dualistic and discriminative thinking, I will cultivate openness, nondiscrimination, and nonattachment to views in order to transform violence, fanaticism, and dogmatism in myself and in the world.

True Happiness

Aware of the suffering caused by exploitation, social injustice, stealing, and oppression, I am committed to practicing generosity in my thinking, speaking, and acting. I am determined not to steal and not to possess anything that should belong to others; and I will share my time, energy, and material resources with those who are in need. I will practice looking deeply to see that the happiness and suffering of others are not separate from my own happiness and suffering; that true happiness is not possible without understanding and compassion; and that running after wealth, fame, power, and sensual pleasures can bring much suffering and despair. I am aware that happiness depends on my mental attitude and not on external conditions, and that I can live happily in the present moment simply by remembering that I already have more than enough conditions to be happy. I am committed to practicing right livelihood so that I can help reduce the suffering of living beings on earth and reverse the process of global warming.

True Love

Aware of the suffering caused by sexual misconduct, I am committed to cultivating responsibility and learning ways to protect the safety and integrity of individuals, couples, families, and society. Knowing that sexual desire is not love, and that sexual activity motivated by craving always harms myself as well as others, I am determined not to engage in sexual relations without true love and a deep, long-term commitment made known to my family and friends. I will do everything in my power to protect children from sexual abuse and to prevent couples and families from being broken by sexual misconduct. Seeing that body and mind are one, I am committed to learning appropriate ways to take care of my sexual energy and to cultivating loving-kindness, compassion, joy, and inclusiveness—which are the four basic elements of true love—for my greater happiness and the greater happiness of others. Practicing true love, we know that we will continue beautifully into the future.

Loving Speech and Deep Listening

Aware of the suffering caused by unmindful speech and the inability to listen to others, I am committed to cultivating loving speech and compassionate listening in order to relieve suffering and to promote reconciliation and peace in myself and among other people, ethnic and religious groups, and nations. Knowing that words can create happiness or suffering, I am committed to speaking truthfully, using words that inspire confidence, joy, and hope. When anger is becoming manifest in me, I am determined not to speak. I will practice mindful breathing and walking in order to recognize and to look deeply into my anger. I know that the roots of anger can be found in my wrong perceptions and lack of understanding of the suffering in myself and in the other person. I will speak and listen in a way that can help myself and the other person transform suffering and see the way out of difficult situations. I am determined not to spread news that I do not know to be certain and not to utter words that can cause division or discord. I will practice right diligence to nourish my capacity for understanding, love, joy, and inclusiveness, and gradually transform the anger, violence, and fear that lie deep in my consciousness.

Nourishment and Healing

Aware of the suffering caused by unmindful consumption, I am committed to cultivating good health, both physical and mental, for myself, my family, and my society by practicing mindful eating, drinking, and consuming. I will practice looking deeply into how I consume the four kinds of nutriments—namely, edible foods, sense impressions, volition, and consciousness. I am determined not to gamble or to use alcohol, drugs, or any other products that contain toxins, such as certain Web sites, electronic games, TV programs, films, magazines, books, and conversations. I will practice coming back to the present moment to be in touch with the refreshing, healing, and nourishing elements in me and around me, not letting regrets and sorrow drag me back into the past or letting anxieties, fears, or cravings pull me out of the present moment. I am determined not to try to cover up loneliness, anxiety, or other suffering by losing myself in consumption. I will contemplate interbeing and consume in a way that preserves peace, joy, and well-being in my body and consciousness, and in the collective body and consciousness of my family, my society, and the earth.

For a full commentary on the Five Mindfulness Trainings, see *For a Future to Be Possible,* by Thich Nhat Hanh. Source: Plum Village Web site, http://www.plumvillage.org/mindfulness-trainings/3-the-five-mindfulness-trainings.html.

Reach the Point of No Failure

Now you have all the tools you need to cultivate mindful living and embark on your mindful journey to attain a healthier weight. Bear in mind that having faith that you can accomplish your intended goal and being diligent in your practice are keys to reaching your destination. Doing everything the way you most enjoy doing it helps you remain mindful. With mindfulness practice, there is no failure—only awareness of what works, what doesn't work, and how to improve the practice at hand. Every mindful step counts. The important thing on this journey is the process, not the destination. Throughout the day, there are many opportunities available to cultivate mindfulness. Find the daily tasks, chores, and events that inspire you the most for integrating mindfulness.

Surrender to what life presents to you in the moment. Stay open, and allow events to unfold. Bear witness to those events, and observe them with equanimity. What is that particular event trying to teach you or alert you to? Stay concentrated on your intentions and goals. With mindfulness, concentration, and your ability to remain nonjudgmental, you will keep gaining valuable insights into all that is needed for you to make progress. This cultivation also helps you become more solid on your path and freer from worries and anxieties over failure.

Through your daily reflection with your mindful living log, thank all the people, beings, and things that have made your life and the way you live possible, as well as those who have helped you stay on the path of mindful living. For example, think of the salad that you had for lunch: how many people and processes are needed to assemble the rainbow-colored salad that you ate? Such contemplation will reinforce your awareness of how blessed and supported you are every day by people you know and by many people you do not know. Thinking like this, you will realize that life is a miracle, that we have much to be grateful for, and that we are all very lucky to have the opportunity to savor it all.

PART 3

Individual and Collective Effort

CHAPTER 8

A Mindful World

There are hundred thousands of stems linking us to
everything in the cosmos,
Supporting us and making it possible for us to be.
Do you see the link between you and me?
If you are not there,
I am not here.

—Thich Nhat Hanh, *The Heart of Understanding* (1988)

WE HAVE PRESENTED THE latest science-based advice, and the age-old tradition of mindfulness, as means to help us understand and transform the unhealthy habits that have led to our overweight. With consistent practice, you will be well on your way to adhering to lifestyle choices that improve your well-being. As you practice mindful living in your daily life, you will gain further insight into the interconnectedness of all things and begin to see the effect of your daily practice extending well beyond your own body—and beyond your own well-being. Science informs us that when our diet is more plant based, and when we exercise regularly, our health will improve. By mindfully reducing meat consumption, you are also performing a miracle, because your change in diet indirectly helps make food more available to hungry children in underdeveloped countries as well as reduce global warming. When more of us practice mindfulness this way, we are creating transformation at not only the individual level but also the collective level. We are changing the world.

Our well-being and the well-being of the world are mutually dependent. We need to stay well at the individual level, and then we will be able to contribute to the well-being of others. By living mindfully, and seizing every moment to live with understanding and compassion, we improve not only our own health but that of all future generations.

The Wise Graduate

On her graduation day in June 2007, senior and salutatorian Jennifer Leigh Levye delivered her address to her Sharon High School classmates in Sharon, Massachusetts. In her speech, she cited the teachings of Thich Nhat Hanh, which were passed on to her by her sophomore English teacher, Mr. Murray.

> Thich Nhat Hanh, . . . holding up an orange, said, "The entire world is inside this orange." "Inside an orange?" was the thought that ran through many of our minds. How can the entire world be inside something as small as an orange?
>
> But think about everything that goes into making an orange. The tree it grows on, the ground it grows in, the water that nourishes it, the sun that gives it the energy to grow. If any of these things are removed, the orange won't exist, so they are all inside the orange. That idea can be pulled out further—a person had to plant the tree, another probably picked the fruit, a third packaged and shipped it. Each of these people was influenced by others and so on until everyone can be connected to the orange—if any one thing were different, it would not exist.

So, from . . . that orange, we see both that our ideas and experiences intertwine and that we are all linked to each other. This isn't the most difficult of ideas to talk about, but it seems very difficult to put into practice. We tend to see ourselves as isolated islands, or as too insignificant to have an effect on the world as a whole. We often think of other people's problems as remote and having no impact on ourselves. We believe that foreign ideas are separate from ours, and we do not really need to understand them. Though these assumptions are common and easy to fall back on, we can remember that our actions do not occur in bubbles, and that we are impacted by the lives of people outside our little spheres. If the entire world is inside an orange, it is also inside each and every one of us—if any part of it were different, we would be different.

. . . [R]emember that everything is connected and that the whole world is inside us. Everyone influences us in a great web, and we can influence it by changing some small part. . . . [I]f we take away from these four years the idea that we are all connected and that we can bring about change, our education has not been in vain, and the world will become a better place."[1]

It is impressive that a young woman like Jennifer already had the wisdom to realize that we do not exist as separate islands. Everything in the universe depends on everything else for its existence. There is nothing that can exist as a separate, independent entity. We are all connected. Our thoughts, speech, and actions affects our family, our community, and our country, rippling out to the entire world. At the same time, the state of our family, community, country, and world affects our own state of being. When Jennifer said, "Everyone influences us in a great web, and we can influence it by changing some small part," she raised the awareness of all her classmates and called them to action.

Individual and Collective Well-Being

Self and *other* are concepts created by us as conventional notions to communicate our ordinary, everyday perceptions. Although these concepts can facilitate communication, they often mislead, and cloud our understanding of the true nature of reality. In our ordinary perception, we see things as independent of one another. We generally perceive things according to preexisting constructs in our mind, which manifest themselves from seeds in our store consciousness. If we are not mindful, these mind constructs may distort the true reality of what we experience. We are often caught and misled by these conventional concepts and their many illusions of duality, like self and other, you and me, inside and outside, being and nonbeing, coming and going, individual and collective, one and many, life and death.

In our ordinary perception, we readily understand a table as a separate, independent object with a flat surface and four legs. But if we look deeply into the table, we see that it is made only of non-table elements—all the phenomena needed for its manifestation: wood, earth, water, fire, air, space, and time. The existence of the table depends on the causes and conditions of all these elements in the entire universe. That is the interdependent nature of the table. It cannot come into being if any one of the conditions or elements is missing.

This is *interbeing*. One thing depends on the manifestation of all other things, and what makes the all possible is the one. One is all, and all is one. In the one you touch the all, and in the all you touch the one. Everything in the universe is present in each of us. I am in you, and you are in me.

I know that you are still there because I am still here.
The arms of perception embrace all,
joining life to death, subject to object, everything to everything else.[2]

We need to cut through the ideas of "individual" and "collective," "inside" and "outside" to see the truth. Inside is made of outside. Our body is not just something that is bound by our skin. Our body is much greater; it is without boundary. For the body to function, we need earth, water, air, heat, and minerals, which are both inside and outside our body. Try to experience the magic of your boundless body the next time you go into a lake or the ocean. Close your eyes, and feel the intimate communion between your body and the water: your body is the water, and the water is your body. The ocean is expansively connected to all things, and so are you.

Looking deeply this way, we can appreciate that the sun is our second heart. If the heart inside our body stops beating, we will die. Similarly, if the sun, our second heart, stops shining, we will also die. Our body is the whole universe, and the whole universe is our body. There is nothing in the universe that is not part of us, be it a speck of dust on the table or a shining star in the sky. This insight is possible only when we transcend the notions of inside and outside, self and other. It is important for us to live mindfully so that we are truly present in every moment, always alive and nourishing the insight of interbeing.

When we practice mindful breathing, smiling, and walking in a group or at a retreat, the collective energy generated by the group helps us build up our own mindfulness energy. When we make an effort to cultivate our mindfulness energy individually in this group setting, we also enhance the energy of the entire group. The individual and the collective are not separate entities. When we touch our inner peace, we smile with joy. The moment we beam a smile, not only do we feel a little happier, but those around us also begin to feel lighter. Individual action always has an effect on the collective, and collective action always has an effect on the individual. When we are looking deeply, the moment we take a mindful step the world changes; everything changes.

When we see more people living mindfully, practicing loving-kindness with understanding and compassion, we gain confidence in our future. When we practice mindful breathing, smiling, eating, walking, and working, we become a positive element in society, and we will inspire confidence in everyone around us. This is the best way to ensure that a future is possible for the younger generations.

From Compassion to Action

To effectively transform our world, we need to touch the source of real strength to mobilize us. Intellect alone cannot motivate us to act compassionately. This strength lies not in power or money, but in our deep, inner peace. This means that we must transform ourselves first to become solid and peaceful. True transformation comes from within. We make changes in our daily lives—in the way we think, speak, act—becoming solid and peaceful, transforming ourselves and the world.

Compassion is a source of powerful, boundless, and wise energy. This is the energy that will move us to act. However, just feeling compassion is not enough; we have to act on it. Understanding and insight guide us to act. That is why love and compassion must always be combined with understanding.

We can cultivate compassion by performing even the smallest acts. If while practicing walking meditation we come across an ant on our path and step aside to avoid crushing it, we are cultivating compassion. If we practice looking deeply and live our daily lives mindfully, our compassion will grow stronger each day. We as individuals can initiate change. When more of us are practicing mindfulness, there will be a change in our collective consciousness. We need to wake ourselves up, and we also need to wake up the collective community. Mindfulness practice at the individual and collective level is the key to this awakening. Our efforts both to change ourselves and to change the environment are necessary, but one cannot happen without the other.

When you live mindfully, taking care of yourself, becoming solid, peaceful, whole, and well, you are empowered to do your part to improve the well-being of all those around you and the well-being of the world. If each of us builds up this collective compassion, we can create a sea change. We must not waste time, but immerse ourselves in the present moment in order to have a clear perspective of our difficult circumstances and transform them.

The practice of mindful consumption should become a global practice. We have to encourage it at both the individual and the collective levels. We need to introduce and support greater mindfulness in all aspects of our lives and invite everyone to join us: parents, educators, students, physicians, social workers, lawyers, scientists, novelists, reporters, filmmakers, businesspeople, architects, artists, farmers, police officers, factory workers, janitors, economists, legislators, and world leaders. This is true peace education.

The World and You

There are many examples that constantly remind us that we are all connected, that we live a life of interbeing. Our tendency is to focus on ourselves for our survival and achievement. Yet in reality we are not separate from others. The financial crisis in the United States in 2008 revealed that we are all interconnected—rich and poor people, farming and manufacturing industries, developed and underdeveloped countries. The banking crisis not only affected the economy in the United States; it also quickly rippled out to affect other markets around the world. The pandemic caused by the H1N1 influenza—often referred to as the "swine" flu—is another concrete example, one in which the source of the virus could well have been pigs. Once one child in Mexico got infected with the virus, the virus quickly spread to Mexico's neighbor, to all of the fifty states of the United States, and within months to forty countries around the world in 2009. With technological advances in communication and travel in the twenty-first century, our world is shrinking, and we are more connected than ever before.

When we are mindfully aware from moment to moment, we realize, "*One contains all. All contains one.*" Just pause to focus on anything that is in front of you—a flower, a computer, a photograph, a glass of water—and you'll see the miracles that connect our world, how all the things we need in our daily life, anywhere and anytime, exist only because of this limitless web of connections. Think of your commute by train or bus from home to work. Could you have access to this transportation without countless others? The train you are riding on is the culmination of years of dedicated hard work by designers, engineers, and craftsmen. The road your bus travels exists only because of the hard work of the highway crew, urban planners, and the many others who also use it. Without all these people, it would take you hours or even an entire day to get to work, just as it does for people who live in developing countries, many of whom still walk on foot to get from one place to another and to meet their daily needs. Thinking this way, you will realize how blessed we are to be able to rely on so many others. We take many things for granted. Many of us tend to be self-centered, living fearfully, with a scarcity mind-set, and trying to accumulate more wealth, power, or status so that we can become more secure. Living like this prevents us from seeing the many wonders that already surround us. Every time we leave our home and commute to work, we can give thanks to all the people who built the commuter trains and buses and constructed the roadways.

There is a human need for meaning, for purposeful connection, for community, and for real engagement in the world. All of us have a great capacity for compassion. We want to help those who are really in need, who are suffering. We want to make the world a better place for this generation and many generations to come. But how do we begin to do this?

Transforming the world starts with oneself. It is through attending to our own well-being and staying in touch with what is happening in our own personal lives that we can have a greater capacity to understand and address the world's suffering. We are then on a sturdier foundation to contribute to improving our world. Have you ever suffered from fatigue or exhaustion from helping others—family, friends, work colleagues, others in your community? Recognize that it is not really possible to steadily help others when we ourselves are not in a good physical, mental, or emotional state. We may be able to carry on for a while, but sooner or later we end up feeling depleted, discouraged, or weak. We cannot keep on giving when we are running on an empty tank. We need to be solid. We need to practice mindful living to be able to offer our best to our family, our friends, and our world.

Influence and Support

For us to be able to eat healthfully and stay physically active, we not only need to have the knowledge, the focus, and the motivation to carry out these daily healthy practices; we also need support from the people and places that we encounter each day—from our closest family and friends, our homes and offices, to the world at large. The Healthy Eating and Active Living Web highlights some forces that can affect our intended goals and the key elements needed for all of us to make healthy eating and active living a way of life. (See figure 8.1.)

As you can see from the illustration, your social environment—friends and family who share your commitment to healthy eating and active living, work colleagues who understand and respect you for taking time to eat a nourishing lunch or go on a walk during your break—has a strong influence on your daily ability to eat and move mindfully. You also need support from your physical environment: If you live in a neighborhood that is unsafe, without adequate lighting, sidewalks, bike paths, parks, or trails for you to bike or walk on, chances are that you will find it much more challenging for you and your family to bike, walk briskly, or jog at will. By the same token, if you live in a "food desert"—a neighborhood that does not have a supermarket—you may have to rely on convenience stores or fast-food restaurants for your food and drink. So instead of being able to easily buy fresh fruits and vegetables and whole grains, you may have to choose your meals from the highly processed, refined carbohydrate-rich foods and the high-sugar drinks that

Figure 8.1 THE HEALTHY EATING AND ACTIVE LIVING WEB

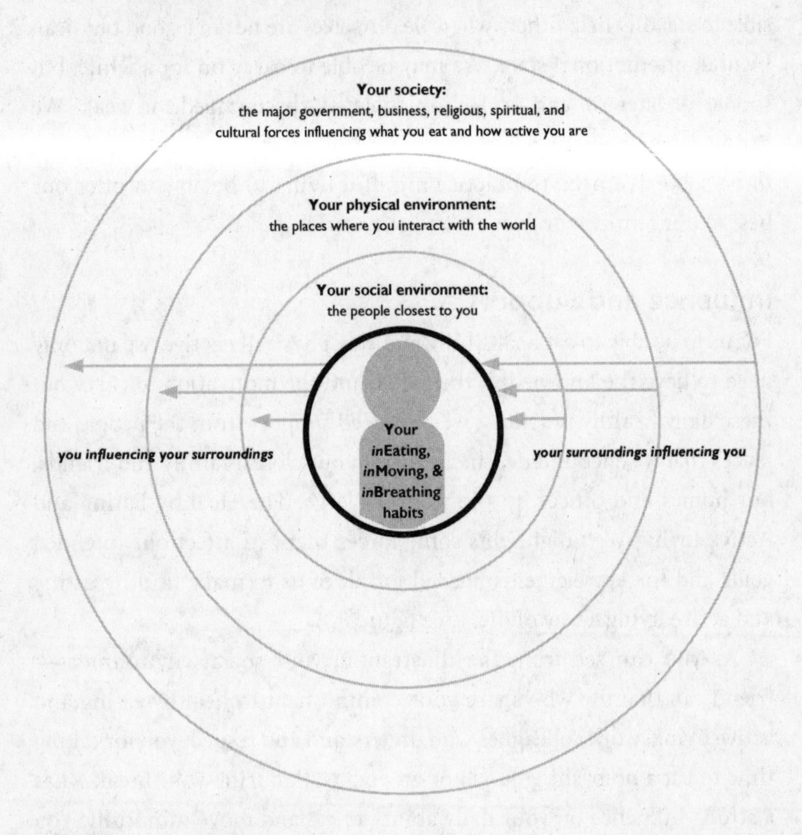

line the convenience-store shelves, or the mega-portioned "meal deals" offered by the fast-food restaurants.

Societal forces also influence the healthy-eating, active-living web. The food industry, ever seeking more profit, controls what foods are offered in the marketplace. The advertising and media industries shape how we view these foods. Agricultural policies influence what foods are most profitable for the food industry to grow. A host of local and national policies also shape our access to physical activity and our environment—architectural and zoning policies, recreation and transportation policies. We need support from all levels of the healthy-eating, active-living web in order for us to change our daily habits for the better. This perspective is described in greater detail in the U.S. Centers for Disease Control and Prevention's recommendation for community strategies to prevent obesity in the United States.[3] A panel of experts identified twenty-four strategies that can help communities create environments that promote healthy eating and active living—among them, improving access to supermarkets in underserved areas, offering incentives to produce food at local farms, improving public transportation, and enhancing personal safety in places where people would be physically active. (The complete article is available at http://www.cdc.gov/mmwr/preview/mmwrhtml/rr5807a1.htm.)

Note that the arrows in figure 8.1 go from the outer rings of the web to the center as well as from the center to the outer rings. This is a reminder that we are influenced by our environment at many levels but that we can also influence our environment.

Be an Agent of Change

As you look deeply at the Healthy Eating and Active Living Web, the longer you contemplate it, the more you realize how many people, businesses, organizations, and governments—at the local, state, and national levels as well as globally—shape your ability to follow a healthy-eating and active-living routine on a daily basis. For many of us, healthy choices are not yet easy choices. When any of these forces are not in line with our healthy-eating or active-living goals, barriers and challenges arise. As Dr. Barry Popkin recounts in his book *The World Is Fat,* technological changes, globalization, government policies, and the food industry have changed our diet and activity over the latter half of the twentieth century. The culmination of all these changes: an unprecedented era of unhealthy eating and sedentary living unlike anything seen previously in our world, in which the number of people who are overweight—1.6 billion—is more than double the number of people who are undernourished.[4] This dire state simply cannot continue, given the economic and health burdens of obesity on individuals, businesses, and countries. All of us can take part in making our world healthier for many generations to come.

This may seem like an overwhelming, daunting, and impossible task, but each one of us can be an agent of change. Here are some examples of inspiring grassroots efforts at social change led by individuals or small groups of people.

From a Mother's Tragedy to a Worldwide Campaign

After the death of her thirteen-year-old daughter to a repeat-offender drunk driver in 1980, Candace Lightner started a small California grassroots organization that became Mothers Against Drunk Driving (MADD), to address the problem of drunk driving.[5] Instead of becoming immobilized by grief, she channeled her grief and anger toward social action. Since then, MADD has become an international organization with more than four hundred chapters worldwide, dealing with drunk driving on multiple levels, from working with city council task forces and state-level legislation to working with a presidential commission on drunk driving. MADD's efforts ultimately led to changes in alcohol policy across all fifty states, including the raising of the legal drinking age and the lowering of the drunk-driving blood-alcohol limit. Lightner's story shows the power that an individual has to make a difference in the world.

From Scientific Inquiry to a Healthier Food Supply

The successful minimization of trans fats in the U.S. food supply at the dawn of the twenty-first century was initiated by the actions of a small group of dedicated scientists. Chemists in the late nineteenth century found they could change liquid vegetable oil into a solid form by adding hydrogen atoms, a process called "partial hydrogenation" that changes the healthy liquid oil into what is known as trans fat.[6] For food manufacturers, this solid form is much more desirable for use in different types of foods such as baked products and margarine, because it is more stable. It does not become rancid easily like liquid vegetable oils, thereby prolonging shelf life. Up until the early 1990s, most people thought that they were reducing their risk of heart disease as long as they stayed away from animal fats, as recommended by health professionals. Margarine with partially hydrogenated oil was considered to be heart healthy, since it does not contain cholesterol. This fact was challenged as research emerged in 1990, when a landmark study by Dr. Martin Katan and his colleagues published in the *New England Journal of Medicine* showed that high levels of trans fat increase LDL ("bad") cholesterol almost as much as saturated fat and, unlike saturated fat, decrease HDL ("good") cholesterol. Since then, this finding has been replicated many times. Subsequently, in 1993 researchers in the Nurses' Health Studies at Harvard University found that trans fat is associated with coronary heart disease in women. These scientific findings provided the evidence for the Center for Science in the Public Interest, a public nutrition advocacy group in the United States, to file a petition requesting that the Food and Drug Administration require the inclusion of trans fats in the nutrition labels of food products. Then in 1994, a research team at the Harvard School of Public Health, led by Dr. Walter Willett, surprised consumers when it found that trans fats are more damaging than saturated fats and are likely to be responsible for at least thirty thousand premature deaths each year in the United States. With this finding, the major international food conglomerate Unilever started to eliminate trans fat from margarine and spreads, which required restructuring at numerous levels including farming, processing, labeling, and advertising.

Meanwhile, scientists continued to publish new findings about the adverse effects of trans fats, pushing for the U.S. Dietary Guidelines and Institute of Medicine reports to include the adverse effects of trans fats, while consumer groups continued to advocate trans-fat labeling on food packages. Finally, the FDA ruled that as of January 1, 2006, trans fat must be listed on food labels along with other bad fats (saturated fats) and good ones (unsaturated fats). In the wake of this ruling and the addition of one line on the food label, food manufacturers and fast-food companies have been overhauling all of their products to reduce or eliminate trans fat. Numerous cities have now declared that restaurants must go trans-fat free, including New York City, Boston, and Philadelphia, reaching what the Center for Science in the Public Interest (CSPI) estimates to be roughly 20 percent of the U.S. population.[7] CSPI also estimates that the amount of trans fat in the U.S. food supply has declined by 50 percent since 2005. These sweeping changes came about because of the dedicated effort of a small group of researchers on two continents working for nearly twenty years, as well as persistent grassroots advocacy by health and consumer education groups.

Sounding the Alarm on Sugary Drinks

We are far from reaching the public-health goal of getting everyone to choose healthier beverages, but the social movement has begun. The movement was ignited by the first scientific study indicating that sugar-sweetened beverages increased the risk of obesity. The 2001 study was based on the work led by Dr. David Ludwig and his colleagues at Harvard University and reported that drinking an additional can of soda a day increased teens' risk of becoming obese by 60 percent.[8] This study, which was widely reported in the press, spurred parents all over the United States to speak up against sugar-sweetened beverages in vending machines, especially in elementary schools. Under pressure, the beverage industries removed sugar-sweetened sodas in all their school contracts with elementary schools in 2006.[9]

Meanwhile, other studies reported increased risks of obesity, diabetes, and heart disease among adults.[10] These scientific studies provided further evidence for the need to control the public's consumption of sugar-sweetened beverages across the entire life span. Insights from the success of the tobacco-control movement in the United States suggest that multiple comprehensive strategies need to be engaged. These include clinical intervention so that doctors can help people lose weight; economic approaches such as taxing sugary drinks; regulatory efforts such as limiting access to sodas and sports drinks in schools; and educational strategies in schools, at work sites, and in the media as well as at point of purchase.[11] We need to work to change the norms around sugary drinks: Water should be the choice nearly all the time, and sugar-sweetened soda should revert back to the "treat" status, something to have once a while, as it was up until a few decades ago.

From a Life Cut Short, Countless Lives Are Saved

In 1986, Countess Albina du Boisrouvray lost her only child, François-Xavier, who was twenty-four years old, during a helicopter mission in Mali.[12] Instead of feeling devastated and drowning in deep sorrow, she decided to devote all her energy and resources to humanitarian causes that perpetuate the values of generosity and compassion that guided her son's life. She founded the Association François-Xavier Bagnoud (FXB International: www.fxb.org) by selling most of her own possessions and getting her friends involved. The mission of FXB, an international, non-governmental organization, is to fight poverty and AIDS and to support orphans and vulnerable children left in the wake of the AIDS pandemic. Understanding the importance of the individual and collective contributions to a community, FXB offers comprehensive support to the families and communities that care for these children, assisting in the planning and building of these communities to be self-sustaining. FXB has over one hundred model programs in Africa, the Americas, Asia, and Europe.

Three Mindful Farmers

Community Supported Agriculture (CSA) is a farm model in which members of the surrounding communities buy shares in a farm; in return, the members get produce and grains grown on that farm. Apart from the exchange of money for fresh, locally grown, often organic products, the CSA model builds important relationships between the communities, the food, and the environment.

There is a notable CSA farm in Santa Cruz, California—Freewheelin' Farm (www.freewheelinfarm.com)—that was started by three young farmers with a vision to provide the community with fresh, organically grown fruits and vegetables while caring for the earth, water, environment, and people that make it all possible. Besides practicing organic farming, they are committed to innovative irrigation and resource consumption analysis, striving to nurture their crops in the most sustainable and environmentally friendly ways. They work to minimize petroleum consumption, and they use recycled materials for all aspects of the farm. As their name suggests, they express their commitment to a greener environment by delivering fresh produce by bicycle!

Beyond growing fresh, healthy food, Freewheelin' Farm is mindful of the need to enhance a collective awareness in the local communities with regard to food, the environment, health, community renewal, and social responsibility, especially in the younger generation. They have developed programs to capture the interests and participation of young people and have been partnering with a youth-empowerment program in the school system to offer hands-on farm training, sustainable-agriculture education, and nutrition education.

The vision of these three farmers is exemplary, and their overall perspective on how to contribute to building a healthier community is admirable. They clearly see the interdependent nature of the land, the food, and the community, and the importance of engaging young people in planting the seeds for our future. This is a wonderful model that should be replicated in communities throughout the world.

Change the World, Step-by-Step

Living in a healthy and supportive environment is a key part of attaining a healthy weight. You may be wondering: What can I do to support the healthy-eating and healthy-living movement? A good place to start is in your immediate environment. What elements are missing from your home environment, your workplace, and your local community that prevent you from nurturing your body and practicing healthy eating and active living? Chances are that the issues we raised in chapters 5 and 6 have alerted you to the many challenges and barriers you face. By now, you realize that you can't blame yourself solely for your past failures to reach your healthy weight. There were many external forces shaping your behavior and preventing you from reaching your goals. Reflect deeply on these challenges and barriers, because understanding them is important in helping you reach your healthy weight and live a mindful life. Everything can begin with you. You are the foundation of any change that will happen in your society. A student asked Thay, "There are so many urgent problems. What should I do?" Thay answered calmly, "Take one thing, and do it very deeply and carefully, and you will be doing everything at the same time."

Knowing that sugar-sweetened beverages are a foe to our weight goals, what can we do to control their consumption? Many of us may say that we just need willpower—that we just need to decide to not drink these beverages. But willpower is likely to be short-lived, especially when we are immersed in an environment that provides easy access to sweet, supersize drinks of twenty ounces a bottle or more. On top of that, we are bombarded with advertisements enticing us to gulp down these drinks. Take action so that you can safeguard yourself from these unhealthy choices.

Start with your immediate environment, your home. Don't keep sugar-sweetened beverages in your fridge. Instead, reconnect to the refreshing taste of tea or water flavored with lime, lemon, orange slices, or mint. When you are thirsty, drink water mindfully and you will discover how refreshing it is. At the cafeteria at work, ask the food-service manager to stock healthier drinks of flavored water and ones that are much lower in sugar—up to three teaspoons per twelve ounces, as recommended by researchers at the Harvard School of Public Health.[13] Do the same for your neighborhood convenience store. Petition, call, or e-mail the consumer hotlines at all the major beverage companies such as Coca-Cola, PepsiCo, and Schweppes or the manufacturer of any of your favorite drinks and ask them to create a lower-sugar drink of up to three teaspoons of sugar per twelve ounces. All businesses are consumer driven. They respond to customer demands. The more consumer requests the beverage industries receive, the more likely it is that they will commit to introducing new products.

You really enjoy riding the bike for your errands around town, but you are worried about safety, since the streets are busy with traffic and there is no bike path. Speak up at your local town meetings, and ask for bike paths to be built. You have already paid taxes to your town, and you have the right to request a safe environment for routine activities. Get your neighbors and friends to do the same, and be persistent. Don't be surprised if your advocacy effort for the bike path encounters some resistance from town officials, who are concerned about costs, or even from some other residents, who may initially see only potential negatives. Find like-minded neighbors to help you get the word out about a bike path's many benefits: A bike path will make families feel more at ease when their children ride their bikes around town, and it will also help reduce your neighborhood's carbon footprint and improve its air quality. Your action will bring the light of mindfulness to your town's elected officials, helping them act on their good intentions for the well-being of all.

Don't underestimate the power of mindful action. You can be a change agent for a healthier environment and for the future. To improve your environment and support healthy eating and active living, there are many needs and options. Take a look at the list of suggestions by the Centers for Disease Control and Prevention in the United States.[14] (See appendix E.) Focus on one area that you are drawn to and passionate about in bringing about change in your community, in your state, or even at the national level. It can be working to bring farmers' markets to your community or revitalizing your community's parks and playgrounds so that children and families can play safely. Or it can be calling the consumer hotlines of food manufacturers and restaurant chains to request healthier choices. There are numerous Web-based resources that give you ideas and help you change the world around you.

To gain more insight and get more ideas on how to promote social change, so that healthy eating and active living can be a reality for all, visit the Web sites of the Center for Science in the Public Interest, in the United States (http://www.cspinet.org); the Rudd Center for Food Policy and Obesity, at Yale University (http://www.yaleruddcenter.org); the Canadian Obesity Network (http://obesitynetwork.ca); Sustain—The Alliance for Better Food and Farming, in the United Kingdom (http://www.sustainweb.org); the Parents Jury, based in Australia (http://www.parentsjury.org.au); Fight the Obesity Epidemic, based in New Zealand (http://www.foe.org.nz); and the International Obesity Task Force (http://www.iotf.org). You can see more ideas for social action at this book's Web site (http://www.savorthebook.com).

In your Mindful Living Plan, include a goal to participate actively in a community-based organization that is focused on improving foods and drinks or your environment. You will be helping not only to improve your community but also to improve your chances of sticking with your own healthy-lifestyle goals as you become surrounded by like-minded volunteers.

Every Day Is a Thanksgiving Day

Every year on the third Thursday of November, Americans celebrate the national holiday of Thanksgiving, which has its roots in the centuries-old Native American tradition of celebrating and giving thanks for the harvest. With our realization of the interdependent nature of *what* and *who* are needed so that we can eat well and stay active every day, Thanksgiving should be celebrated not just once a year. It should be celebrated more frequently—like . . . every day. Make a habit of thanking silently all those whom you are connected to and dependent on, be it your spouse who prepared a healthy, delicious meal; the dedicated researchers who discovered the scientific wisdom on healthy eating; the cashier at the lunch line; the food businesses that offer us the healthy choices we need; the radio reporter announcing the farm bill that has just passed in Congress; the representatives and senators who worked tirelessly to pass legislature to improve our diets; or the physician who encouraged you to work toward a healthier weight. When we are in touch with our gratitude for all of these people, we will be inspired and energized to act and do our part to contribute to improving our own health and quality of life and that of many others.

Savor Every Moment

None of us lives forever. While we are alive, all of us have the choice to live a life of mindfulness, leading to more peace and joy—or to live a life without mindfulness, leading to more anxiety and suffering. Every day, it is important to remind ourselves that all notions, things, and people are impermanent. That sickness and loss are inevitable. And that we need to live in the moment to be truly fulfilled. The Buddha encouraged his students to practice and reflect regularly on these Five Remembrances:

I am of the nature to grow old. There is no way to escape growing old.

I am of the nature to have ill-health. There is no way to escape having ill-heath.

I am of the nature to die. There is no way to escape death.

All that is dear to me and everyone I love are of the nature to change. There is no way to escape being separated from them.

My actions are my only true belongings. I cannot escape the consequences of my actions. My actions are the ground on which I stand.[15]

Savor the time you still have in this life. Savor every moment, every breath, every meal, every relationship, every action or nonaction, every opportunity to maintain your well-being and the well-being of our world. Integrate and practice mindfulness in your everyday living so that it becomes a habit, a way of life. Get others to join you, lending support to each other to eat, work, and live mindfully together. Living like this is your only true belonging and is the essence of a meaningful and fulfilling life.

APPENDIX A

Integrating Mindfulness into Your Daily Life

Mindfulness Practice Centers in the Tradition of Thich Nhat Hanh

Blue Cliff Monastery, 3 Mindfulness Rd., Pine Bush, NY 12566
Phone: (845) 733-4959
E-mail: office@bluecliffmonastery.org
Web site: www.bluecliffmonastery.org

Deer Park Monastery, 2499 Melru Lane, Escondido, CA 92026
Phone: (760) 291-1003
E-mail: deerpark@dpmail.net
Web site: www.deerparkmonastery.org

European Institute of Applied Buddhism, Schaumburgweg 3, D-51545 Waldbröl, Germany
Phone: +49 (0)2291 907 1373
E-mail: info@eiab.eu
Web site: www.eiab.eu

Maison de L'Inspir (House of Breathing, near Paris), 7 Allée des Belles Vues, 93160 Noisy-le-Grand, France
Phone: +33 (0)9 51 35 46 34
E-mail: maisondelinspir@yahoo.fr
Web site: www.maisondelinspir.over-blog.com

ȩe, 13 Martineau, 33580 Dieulivol, France

ɔffice@plumvillage.org; uh-office@plumvillage.org; nh-

llage.org

Web site: www.plumvillage.org

Sangha Directories

Community of Mindful Living. *USA Sangha Directory.* http://www.iamhome.org/usa_sangha.htm

Community of Mindful Living. *International Sangha Directory.* http://www.iamhome.org/international.htm

Mindfulness Clock

Download to your Windows computer the sound of a mindfulness bell that you can program to ring at different intervals, courtesy of the Washington Mindfulness Community: http://www.mindfulnessdc.org/mindfulclock.html.

For Mac computer users, the free downloadable widget ProdMe 1.1 (copyright Jim Carlson), available on the Apple Web site, can be programmed for a "clock chimes" sound and used as a mindfulness bell: http://www.apple.com/downloads/dashboard/status/prodme.html.

The Now Watch

Remind yourself to stay in the here and now with this watch, in Thich Nhat Hanh's own calligraphy, which can be purchased from the bookshop at Blue Cliff Monastery. For more information, e-mail the bookshop at bookshop@bluecliffmonastery.org, or in Europe, e-mail itsnow@plumvillage.org.

APPENDIX B

Discourse on the Four Kinds of Nutriments

Samyukta Agama, sutra 373

This is what I heard one time when the Buddha was in the Anathapindika Monastery in the Jeta Grove near to the town of Shravasti. That day the Buddha told the monks: "There are four kinds of nutriments which enable living beings to grow and maintain life. What are these four nutriments? The first is edible food, the second is the food of sense impressions, the third is the food of volition, and the fourth is the food of consciousness."

"Bhikkhus, how should a practitioner regard edible food? Imagine a young couple with a baby boy whom they look after and raise with all their love. One day they decide to bring their son to another country to make their living. They have to go through the difficulties and dangers of a desert. During the journey, they run out of provisions and fall extremely hungry. There is no way out for them and they discuss the following plan: 'We only have one son whom we love with all our heart. If we eat his flesh we shall survive and manage to overcome this dangerous situation. If we do not eat his flesh all three of us will die.' After this discussion, they killed their son, with tears of pain and gritting their teeth they ate the flesh of their son, just so as to be able to live and come out of the desert."

The Buddha asked: "Do you think that couple ate their son's flesh because they wanted to enjoy its taste and because they wanted their bodies to have the nutriment that would make them more beautiful?"

The monks replied: "No, Venerable Lord." The Buddha asked: "Were the couple forced to eat their son's flesh in order to survive and escape from the dangers of the desert?" The monks replied: "Yes, Venerable Lord."

The Buddha taught: "Monks, every time we ingest edible food, we should train ourselves to look at it as our son's flesh. If we meditate on it in this way we shall have clear insight and understanding which puts an end to misperceptions about edible food and our attachment to sensual pleasures will dissolve. Once the attachment to sensual pleasures is transformed there are no longer any internal formations concerning the five objects of sensual pleasure in the noble disciple who applies himself to the training and the practice. When the internal formations still bind us we have to keep returning to this world.

"How should the practitioner meditate on the food of sense impressions? Imagine a cow which has lost its skin. Wherever it goes the insects and maggots which live in the earth, in the dust and on the vegetation attach themselves to the cow and suck its blood. If the cow lies on the earth, the maggots in the earth will attach themselves to it and feed off of it. Whether lying down or standing up, the cow will be irritated and suffer pain. When you ingest the food of sense impressions, you should practice to see it in this light. You will have insight and understanding which puts an end to misperceptions concerning the food of sense impressions. When you have this insight you will no longer be attached to the three kinds of feeling. When no longer attached to the three kinds of feeling the noble disciple does not need to strive anymore because whatever needs to be done has already been done.

"How should the practitioner meditate on the food of volition? Imagine there is a village or a large town near to a pit of burning charcoal. There are only the smokeless, glowing embers left. Now there is an intelligent man with enough wisdom who does not want to suffer and only wants happiness and peace. He does not want to die and he only wants to live. He thinks: 'Over there the heat is very great, although there is no smoke and there are no flames. Still, if I have to go into that pit there is no doubt that I shall die.' Knowing this he is determined to leave that large town or that village and go somewhere else. The practitioner should meditate like this on the food of volition. Meditating like this he will have insight and understanding which puts an end to misperceptions about the food of volition. When he arrives at that understanding the three kinds of craving will be ended. When these three cravings are ended, the noble disciple who trains and practices will have no more work to do, because whatever needs to be done has already been done.

"How should the practitioner meditate on the food of consciousness? Imagine that the soldiers of the king have arrested a criminal. They bind him and bring him to the king. Because he has committed theft he is punished by people piercing his body with three hundred knives. He is assailed by fear and pain all day and all night. The practitioner should regard the food of consciousness in this light. If he does he will have insight and understanding which puts an end to misperceptions concerning the food of consciousness. When he has this understanding regarding the food of consciousness the noble disciple who trains and practices will not need to strive anymore because whatever needs to be done has been done."

When the Buddha had spoken, the monks were very happy to put the teachings into practice.

APPENDIX C

Total Relaxation

(from *The Art of Power*, by Thich Nhat Hanh)

People in any profession can profit from doing total relaxation every day or every week. Maybe there's a place in the office where you can practice total relaxation for fifteen minutes. You can do it at home also. Once you have benefited from practicing on your own to renew yourself, you can offer a session of total relaxation to your whole family, and to your colleagues at work. You could have a session of total relaxation at your job every day. When coworkers and employees become overwhelmed by stress, they are much less effective in their work and often miss work because of sickness. This is very costly to the organization. So fifteen minutes of total relaxation after three or four hours of work is very practical. You may like to lead the total relaxation yourself. You will experience a lot of joy doing this. When you're able to make people happy and relaxed, your own happiness increases at the same time.

When we do deep relaxation in a group, one person can guide the exercise using the following instructions, or some variation of them. You may want to invite a bell at the beginning and end of the exercise to help people more easily enter a relaxed state of mind. When you do deep relaxation on your own, you can make a recording that you can play to guide you.

Lie down on your back with your arms at your sides. Make yourself comfortable. Allow your body to relax. Be aware of the floor underneath you . . . and of the contact of your body with the floor. Allow your body to sink into the floor.

Become aware of your breathing, in and out. Be aware of your abdomen rising and falling as you breathe in and out . . . rising . . . falling . . . rising . . . falling.

Breathing in, breathing out . . . your whole body feels light . . . like a water lily floating on the water . . . You have nowhere to go . . . nothing to do . . . You are free as the cloud floating in the sky.

Breathing in, bring your awareness to your eyes. Breathing out, allow your eyes to relax. Allow your eyes to sink back into your head . . . let go of the tension in all the tiny muscles around your eyes . . . our eyes allow us to see a paradise of forms and colors . . . allow your eyes to rest . . . send love and gratitude to your eyes.

Breathing in, bring your awareness to your mouth. Breathing out, allow your mouth to relax. Release the tension around your mouth . . . your lips are the petals of a flower . . . let a gentle smile bloom on your lips . . . smiling releases the tension in the hundreds of muscles in your face . . . feel the tension release in your cheeks . . . your jaw . . . your throat.

Breathing in, bring your awareness to your shoulders. Breathing out, allow your shoulders to relax. Let them sink into the floor . . . let all the accumulated tension flow into the floor . . . we carry so much with our shoulders . . . now let them relax as we care for our shoulders.

Breathing in, become aware of your arms. Breathing out, relax your arms. Let your arms sink into the floor . . . your upper arms . . . your elbows . . . your lower arms . . . your wrists . . . hands . . . fingers . . . all the tiny muscles . . . move your fingers a little if you need to, to help the muscles relax.

Breathing in, bring your awareness to your heart. Breathing out, allow your heart to relax. We have neglected our heart for a long time by the way we work, eat, and deal with anxiety and stress. Our heart beats for us night and day. Embrace your heart with mindfulness and tenderness, reconciling with and taking care of your heart.

Breathing in, bring your awareness to your legs. Breathing out, allow your legs to relax. Release all the tension in your legs . . . your thighs . . .

your knees . . . your calves . . . your ankles . . . your feet . . . your toes . . . all the tiny muscles in your toes . . . you may want to move your toes a little to help them relax . . . send your love and care to your toes.

Bring your awareness back to your breathing . . . to your abdomen rising and falling.

Following your breathing, become aware of your arms and legs . . . you may want to move them a little and stretch.

When you feel ready, slowly sit up. When you are ready, slowly stand up.

The above exercises can guide awareness to any part of the body: the hair, brain, ears, neck, lungs, each of the internal organs, the digestive system, or to any part of the body that needs healing and attention. You embrace each part and send it your love, gratitude, and care as you hold it in your awareness and breathe in and out.

APPENDIX D

Screen-Time Alternatives

- Acting
- Bike riding
- Bowling
- Camping
- Caring for your pet
- Cleaning the house
- Cooking
- Crafting
- Dancing
- Doing laundry
- Doing push-ups
- Doing sit-ups
- Doing yoga
- Drawing
- Fishing
- Gardening
- Going to a house of worship
- Going to a spiritual center
- Going to the gym
- Golfing
- Grocery shopping
- Hiking
- In-line skating
- Inviting friends over
- Journaling
- Jogging
- Jumping rope
- Knitting or crocheting
- Learning a language
- Listening to music
- Meditating
- Mindful walking
- Mowing the lawn
- Organizing
- Painting
- Planting flowers
- Playing an instrument
- Playing board games
- Playing *Dance, Dance Revolution*
- Playing Frisbee
- Playing tag
- Playing team sports
- Playing tennis
- Playing with a hacky sack
- Playing with children
- Reading
- Relaxing
- Scrapbooking

- Singing
- Sleeping
- Solving number puzzles
- Spending time with family
- Storytelling
- Stretching
- Swimming
- Taking photographs
- Talking on the phone with friends
- Tutoring
- Visiting a museum
- Volunteering
- Walking
- Woodworking
- Writing poetry

Adapted from Lilian W. Y. Cheung, Hank Dart, Sari Kalin, and Steven L. Gortmaker, *Eat Well & Keep Moving*, 2nd ed. (Champaign, IL: Human Kinetics), p. 387.

APPENDIX E

Community-Based Strategies to Prevent and Control Obesity

Strategies to Promote the Availability of Affordable Healthy Food and Beverages

1. Communities should increase the availability of healthier food and beverage choices in public service venues.
2. Communities should improve the availability of affordable, healthier food and beverage choices in public service venues.
3. Communities should improve the geographic availability of supermarkets in underserved areas.
4. Communities should provide incentives to food retailers to locate in underserved areas and/or to offer healthier food and beverage choices in those areas.
5. Communities should improve the availability of mechanisms for purchasing foods from farms.
6. Communities should provide incentives for the production, distribution, and procurement of foods from local farms.
7. Communities should restrict the availability of less-healthy foods and beverages in public service venues.

Source: L. K. Khan et al., Recommended community strategies and measurements to prevent obesity in the United States, *MMWR Recommendations and Reports* 58 (2009): 1–26. For more information on all of these strategies, see the full article on the journal's Web site: http://www.cdc.gov/mmwr/preview/mmwrhtml/rr5807a1.htm.

8. Communities should institute smaller portion size options in public service venues.
9. Communities should limit advertisements of less-healthy foods and beverages.
10. Communities should discourage the consumption of sugar-sweetened beverages.

Strategy to Encourage Breast-Feeding

Communities should increase support for breast-feeding.

Strategies to Encourage Physical Activity or Limit Sedentary Activity Among Children and Youth

1. Communities should require physical education in schools.
2. Communities should increase the amount of physical activity in physical education programs in schools.
3. Communities should increase the opportunities for extracurricular physical activity.
4. Communities should reduce computer-screen time in public service venues, such as schools, day care centers, and after-school programs.

Strategies to Create Safe Communities That Support Physical Activity

1. Communities should improve access to outdoor recreational facilities.
2. Communities should enhance infrastructure that supports bicycling.
3. Communities should enhance infrastructure that supports walking.
4. Communities should support locating schools within easy walking distance of residential areas.
5. Communities should improve access to public transportation.

6. Communities should zone for mixed-use development.
7. Communities should enhance personal safety in areas where people are or could be physically active.
8. Communities should enhance traffic safety in areas where people are or could be physically active.

Strategy to Encourage Communities to Organize for Change

Communities should participate in community coalitions or partnerships to address obesity.

APPENDIX F

Resources

Healthy Eating and Drinking

Books

Nestle, M. *What to Eat.* New York: North Point Press, 2006.

Willett, W. C., and P. J. Skerrett. *Eat, Drink, and Be Healthy: The Harvard Medical School Guide to Healthy Eating.* New York: Free Press, 2005.

Web Sites and Online Publications

American Institute for Cancer Research. *American Institute for Cancer Research Home Page.* http://www.aicr.org.

American Institute for Cancer Research, World Cancer Research Foundation. *Food, Nutrition, Physical Activity, and the Prevention of Cancer: A Global Perspective—Online.* August 19, 2009; http://www.dietandcancerreport.org/.

Center for Science in the Public Interest. *Nutrition Action Health Letter.* http://www.cspinet.org/nah/index.htm.

Centers for Disease Control and Prevention. *Healthy Eating for a Healthy Weight.* http://www.cdc.gov/healthyweight/healthy_eating/.

Cooper Institute. *Stand Up (More) and Eat (Better).* http://www.standupandeat.org/.

Harvard School of Public Health. *The Nutrition Source: Knowledge for Healthy Eating.* http://www.thenutritionsource.org.

National Institute on Alcohol Abuse and Alcoholism. *College Drinking—Changing the Culture.* http://www.collegedrinkingprevention.gov/.

U.S. Department of Agriculture. *My Pyramid Tracker.* http://www.mypyramidtracker.gov/.

U.S. Department of Health and Human Services, Substance Abuse and Mental Health Services Administration. *Alcohol & Drug Information.* http://ncadi.samhsa.gov/.

U.S. Food and Drug Administration. *Food Labeling and Nutrition Education Tools.* http://www.fda.gov/Food/LabelingNutrition/ConsumerInformation/ucm121642.htm.

Healthy Weight

Books

Critser, G. *Fat Land.* New York: Penguin Books, 2004.

Hu, F. B. *Obesity Epidemiology.* New York: Oxford Univ. Press, 2008.

Institute of Medicine. *Progress in Preventing Childhood Obesity: How Do We Measure Up?* Washington, DC: National Academies Press, 2006.

Katzen, M., and W. C. Willett. *Eat, Drink, & Weigh Less: A Flexible and Delicious Way to Shrink Your Waist Without Going Hungry.* New York: Hyperion, 2006.

Kumanyika, S. *Handbook of Obesity Prevention: A Resource for Health Professionals.* New York: Springer, 2007.

Kushner, R. F. *Dr. Kushner's Personality Type Diet.* New York: St. Martin's Griffin Press, 2004.

Ludwig, D. S., with S. Rostler. *Ending the Food Fight: Guide Your Child to a Healthy Weight in a Fast Food/Fake Food World.* New York: Houghton Mifflin Harcourt, 2007.

Neumark-Sztainer, D. *I'm Like So Fat.* New York: Guilford Publications, 2005.

Popkin, B. M. *The World Is Fat.* New York: Penguin Group, 2009.

Satter, E. *How to Get Your Kid to Eat . . . but Not Too Much.* Boulder: Bull Publishing Company, 1987.

Web Sites and Online Publications

Action for Healthy Kids. *Action for Healthy Kids Home Page.* http://www.actionforhealthykids.org/.

Brown Medical School. *The National Weight Control Registry.* http://www.nwcr.ws/.

California Project LEAN. *California Project LEAN: Leaders Encourage Activity and Nutrition.* http://www.californiaprojectlean.org/.

Canadian Obesity Network. *Canadian Obesity Network Home Page.* http://www.obesitynetwork.ca/.

Centers for Disease Control and Prevention. *The Guide to Community Preventive Services—Obesity Prevention.* http://www.thecommunityguide.org/obesity/index.html.

———. *Overweight and Obesity.* http://www.cdc.gov/obesity/.

———. *Overweight and Obesity: State Stories.* http://www.cdc.gov/obesity/stateprograms/statestories.html.

Centers for Science in the Public Interest. *Nutrition Policy.* http://www.cspinet.org/nutritionpolicy/index.html.

Harvard School of Public Health. *Harvard Prevention Research Center on Nutrition and Physical Activity.* http://www.hsph.harvard.edu/prc/.

International Obesity Task Force. *International Obesity Task Force Home Page.* http://www.iotf.org.

Louisiana State University. *Pennington Biomedical Research Center.* http://www.pbrc.edu/.

Mayo Clinic. *MayoClinic.com—Healthy Lifestyle.* http://www.mayoclinic.com/health/HealthyLivingIndex/HealthyLivingIndex.

Mayo Clinic Staff. *Calorie Calculator.* http://www.mayoclinic.com/health/calorie-calculator/NU00598.

National Eating Disorders Association. *National Eating Disorders Association.* http://www.nationaleatingdisorders.org/.

National Heart Lung and Blood Institute. *Calculate Your Body Mass Index.* http://www.nhlbisupport.com/bmi/.

National Institute of Diabetes and Digestive and Kidney Diseases. *Weight-control Information Network (WIN).* http://win.niddk.nih.gov/.

National Institutes of Health. *We Can! (Ways to Enhance Children's Activity & Nutrition).* http://www.nhlbi.nih.gov/health/public/heart/obesity/wecan/.

New England Coalition for Health Promotion and Disease Prevention. *Strategic Plan for Prevention and Control of Overweight and Obesity in New England.* http://www.neconinfo.org/02–11–2003_Strategic_Plan.pdf.

Robert Wood Johnson Foundation. *Active Living Research.* http://www.activelivingresearch.org/.

———. *Healthy Eating Research.* http://www.healthyeatingresearch.org/.

Shape Up America! *Shape Up America! Healthy Weight for Life.* http://www.shapeup.org/.

Siteman Cancer Center, Washington University School of Medicine. *Your Disease Risk: The Source on Prevention.* http://www.yourdiseaserisk.org/.

Trust for America's Health. *F as in Fat 2009: How Obesity Policies Are Failing in America.* http://healthyamericans.org/reports/obesity2009/.

Tufts University. *Children in Balance.* http://www.childreninbalance.org/.

University of California at Berkeley. *Dr. Robert C. and Veronica Atkins Center for Weight & Health.* http://cwh.berkeley.edu/.

Yale University. *Rudd Center for Food Policy and Obesity.* http://www.yaleruddcenter.org/.

How Food Reaches Our Table

Books

Brownell, K. D., and K. B. Horgen. *Food Fight: The Inside Story of the Food Industry, America's Obesity Crisis, and What We Can Do About It.* New York: McGraw-Hill, 2003.

Kingsolver, B. *Animal, Vegetable, Miracle.* New York: HarperCollins, 2007.

Nestle, M. *Food Politics: How the Food Industry Influences Nutrition and Health.* 2nd ed. California Studies in Food and Culture 3. Berkeley: Univ. of California Press, 2007.

Pollan, M. *In Defense of Food: An Eater's Manifesto.* New York: Penguin, 2008.

———. *The Omnivore's Dilemma: A Natural History of Four Meals.* New York: Penguin, 2006.

Schlosser, E. *Fast Food Nation: The Dark Side of the All-American Meal.* New York: Houghton Mifflin, 2001.

Web Sites and Online Publications

Edible Schoolyard. *The Edible Schoolyard Home Page.* http://www.edibleschoolyard.org/.

FoodRoutes Network. *FoodRoutes.org Home Page.* http://www.foodroutes.org.

Kitchen Gardeners International. *Kitchen Gardeners International Home Page.* http://www.kitchengardeners.org/.

Slow Food. *Slow Food International—Good, Clean, and Fair Food.* http://www.slowfood.com/.

Small Planet Institute. *Small Planet Institute Home Page.* http://smallplanetinstitute.org/.

Sustainable Table. *Sustainable Table: Serving Up Healthy Food Choices.* http://www.sustainabletable.org/home.php.

Movies

Kenner, R. and E. Schlosser. *Food Inc.* 2008. More information available at http://www.foodincmovie.com/.

Spurlock, M. *Super Size Me.* 2004.

Stein, J., *Peaceable Kingdom.* 2004. More information available at http://www.peaceablekingdomfilm.org/.

Woolf, A., *King Corn.* 2007. More information available at http://www.kingcorn.net/.

Making Change in Your Community and in the World

Web Sites and Online Publications

Active Living by Design. *Active Living by Design Home Page.* http://www.activelivingbydesign.org/.

Active Living Resource Center. *Active Living Resource Center Home Page.* http://www.activelivingresources.org/index.php.

Center for Science in the Public Interest. *Center for Science in the Public Interest Home Page.* http://www.cspinet.org/.

Change.org. *Change.org Home Page.* http://www.change.org/.

Fight the Obesity Epidemic. *Fight the Obesity Epidemic: Stop Our Children Developing Type 2 Diabetes.* http://www.FOE.org.NZ.

Free the Children. *Free the Children Home Page.* http://freethechildren.com/.

Harvard School of Public Health. *The Francois-Xavier Bagnoud Center for Health and Human Rights.* http://www.harvardfxbcenter.org/.

National Center for Biking and Walking. *BikeWalk.org.* http://www.bikewalk.org/index.php.

Parents Jury. *The Parents Jury: Your Voice on Food and Activity.* http://www.parentsjury.org.au.

Sustain: The Alliance for Better Food and Farming. *Sustainweb.org.* http://www.sustainweb.org/.

USA.gov. *Contact Elected Officials.* http://www.usa.gov/Contact/Elected.shtml.

Mindful Eating

Books

Albers, S. *Eating Mindfully: How to End Mindless Eating and Enjoy a Balanced Relationship with Food.* Oakland, CA: New Harbinger Publications, 2003.

Bays, J. *Mindful Eating.* Boston: Shambhala, 2009.

Goodall, J. *Harvest for Hope: A Guide to Mindful Eating.* New York: Time Warner Book Group, 2005.

Kessler, D. A. *The End of Overeating: Taking Control of the Insatiable American Appetite.* New York: Rodale Press, 2009.

Yuen, C. *The Cosmos in a Carrot: A Zen Guide to Eating Well.* Berkeley: Parallax Press, 2006.

Web Sites and Online Publications

Center for Mindful Eating. *The Center for Mindful Eating Home Page.* http://www.tcme.org/library.htm.

Wansink, B. *MindlessEating.org.* http://www.mindlesseating.org/.

Mindfulness

Books

Badiner, A. H., ed. *Mindfulness in the Marketplace: Compassionate Responses to Consumerism.* Berkeley: Parallax Press, 2002.

Hanh, T. N. *Anger: Wisdom for Cooling the Flames.* New York: Riverhead Books, 2001.

———. *The Art of Power.* New York: HarperCollins, 2007.

———. *Creating True Peace.* New York: Simon and Schuster, 2003.

———. *The Heart of the Buddha's Teachings.* Berkeley: Parallax Press, 1998.

———. *Peace Is Every Step: The Path of Mindfulness in Everyday Life.* New York: Bantam Books, 1991.

Langer, E. J. *Mindfulness.* Cambridge, MA: Perseus, 1989.

Siegel, D. *The Mindful Brain.* New York: Norton, 2007.

Newsletters and Magazines

Community of Interbeing UK. *Here and Now Newsletter.* http://www.interbeing.org.uk/here&now/.

Mindfulness Bell: A Journal of the Art of Mindful Living. You can subscribe to and learn more about the *Mindfulness Bell* at www.iamhome.org.

Web Sites and Online Publications

Community of Interbeing UK. *The Community of Interbeing Manual of Practice.* http://www.interbeing.org.uk/manual/.

Community of Mindful Living. *International Sangha Directory.* http://www.iamhome.org/international.htm.

———. *USA Sangha Directory.* http://www.iamhome.org/usa_sangha.htm.

Mindful Kids. *Mindful Kids Blog: Sharing Mindfulness with Children.* http://mindfulkids.wordpress.com/.

Plum Village. *Practice: Art of Mindful Living.* http://plumvillage.org/practice.html. The site includes a helpful list of all the mindfulness practices in the Thich Nhat Hanh tradition.

UCLA Semel Institute. *UCLA Mindful Awareness Research Center.* http://marc.ucla.edu/.

University of Massachusetts Medical School. *Center for Mindfulness in Medicine, Health Care, and Society.* http://www.umassmed.edu/content.aspx?id=41252.

University of Pennsylvania. *Authentic Happiness Home Page.* http://www.authentichappiness.sas.upenn.edu/Default.aspx.

Wake Up. *Wake Up: Young Buddhists and Non-Buddhists for a Healthy and Compassionate Society.* http://wkup.org/.

Physical Activity and Exercise

Books

Anderson, B. *Stretching.* Illustrated by J. Anderson. Bolinas, CA: Shelter Publications, 2000.

Boccio, F. J., and G. Feuerstein. *Mindfulness Yoga: The Awakened Union of Breath, Body and Mind.* Somerville, MA: Wisdom Publications, 2004.

Chryssicas, M. K. *Breathe: Yoga for Teens.* New York: DK Publishing, 2007.

Cooper, K. H., and T. C. Cooper. *Start Strong, Finish Strong: Prescriptions for a Lifetimes of Great Health.* New York: Penguin Group, 2007.

Hanh, T., and W. Vriezen. *Mindful Movements.* Berkeley: Parallax Press, 2008.

Komitor, J. B. *The Complete Idiot's Guide to Yoga with Kids.* Indianapolis, IN: Prentice Hall, 2000.

Manson, J., and P. Amend. *The 30-Minute Fitness Solution: A Four-Step Plan for Women of All Ages.* Cambridge, MA: Harvard Univ. Press, 2001.

Nelson, M. A., *Strong Women, Strong Bones.* New York: C.P. Putnam & Sons, 2000.

Ratey, J. J. *Spark: The Revolutionary New Science of Exercise and the Brain.* Boston: Little, Brown & Company, 2008.

Special Reports

Harvard Health Publishing. *Exercise: A Program You Can Live With.* Boston, 2007.

———. *Strength and Power Training: A Guide for Adults of All Ages.* Boston, 2007.

Web Sites and Online Publications

Attack Point. *Attack Point Home Page.* http://attackpoint.org/.

Centers for Disease Control and Prevention. *National Physical Activity Plan.* http://www.physicalactivityplan.org/.

———. *Physical Activity.* http://www.cdc.gov/physicalactivity/.

Gmaps Pedometer. *Gmaps Pedometer Home Page.* http://www.gmap-pedometer.com/.

Mayo Clinic Staff. *Weight Training Exercises for Major Muscle Groups.* http://www.mayoclinic.com/health/weight-training/SM00041.

National Institute on Aging. *Exercise and Physical Activity: Your Everyday Guide from the National Institute of Aging.* http://www.nia.nih.gov/healthinformation/publications/exerciseguide.

President's Council on Physical Fitness and Sports. *Fitness.gov.* http://www.fitness.gov/.

USA Track and Field. *America's Running Routes.* http://www.usatf.org/routes/.

U.S. Department of Health and Human Services. *2008 Physical Activity Guidelines for Americans: Be Active, Healthy, and Happy!* http://www.health.gov/paguidelines/guidelines/default.aspx.

Sleep

Books

Epstein, L. *The Harvard Medical School Guide to a Good Night's Sleep.* New York: McGraw-Hill, 2007.

Web Sites and Online Publications

Harvard Medical School. *Healthy Sleep.* http://healthysleep.med.harvard.edu/.

National Sleep Foundation. *SleepFoundation.org.* http://www.sleepfoundation.org/.

Sustainable Agriculture and Protecting the Planet

Books

Brown, L. R. *Plan B 3.0: Mobilizing to Save Civilization.* New York: W. W. Norton & Company, 2008. Also available for free download at http://www.earthpolicy.org/images/uploads/book_files/pb3book.pdf.

Fukuoka, M. *The One-Straw Revolution: An Introduction to Natural Farming.* New York: The New York Review of Books, 2009.

Hanh, T. N. *The World We Have.* Berkeley: Parallax Press, 2008.

Web Sites and Online Publications

The Climate Project. *TheClimateProject.Org.* http://www.theclimateproject.org.

Deer Park Monastery. *Earth Peace Treaty.* http://www.earthpeacetreaty.org/.

Gore, A. *ClimateCrisis.Net.* http://www.climatecrisis.net/.

Harvard Medical School. *Center for Health and the Global Environment.* http://chge.med.harvard.edu/index.html.

Harvard University. *Harvard University Center for the Environment.* http://environment.harvard.edu/.

———. *Sustainability at Harvard.* http://www.greencampus.harvard.edu/.

Johns Hopkins University. *Center for a Livable Future.* http://www.jhsph.edu/clf/.

Local Harvest. *Local Harvest Home Page.* http://www.localharvest.org/.

Sustainable Agriculture Research and Education. *Sustainable Agriculture and Research Education Home Page.* http://sare.org/.

United Nations Environment Programme and the World Meteorological Organization. *Intergovernmental Panel on Climate Change (IPCC).* http://www.ipcc.ch/.

USDA National Agricultural Library. *Community Supported Agriculture.* September 28, 2009; http://nal.usda.gov/afsic/pubs/csa/csa.shtml.

———. *Sustainable Agriculture: Information Access Tools.* October 14, 2009; http://nal.usda.gov/afsic/pubs/agnic/susag.shtml.

W. K. Kellogg Foundation. *Overview: Food & Fitness.* http://www.wkkf.org/default.aspx?tabid=75&CID=383&NID=61&LanguageID=0.

World Carfree Network. *World Carfree Day.* http://www.worldcarfree.net/wcrd/.

TV Alternatives and Media Awareness

Books

Bennett, S., and R. Bennett. *365 TV-Free Activities You Can Do with Your Child: Plus 50 All-New Bonus Activities.* Avon, MA: Adams Media Corporation, 2002.

Institute of Medicine. *Food Marketing to Children and Youth: Threat or Opportunity.* Washington, DC: National Academies Press, 2006.

Linn, S. *Consuming Kids: The Hostile Takeover of Childhood.* New York: New Press, 2004.

Web Sites and Online Publications

Campaign for a Commercial-Free Childhood. *Campaign for a Commercial-Free Childhood Home Page.* http://www.commercialexploitation.org/.

Center for Media Literacy. *Center for Media Literacy Home Page.* http://www.medialit.org/.

Center for Screen Time Awareness. *ScreenTime.org.* http://www.screentime.org/index.php?option=com_frontpage&Itemid=1.

Kaiser Family Foundation. *Study of Media and Health.* http://www.kff.org/entmedia/index.cfm.

Media Awareness Network. *Dealing with Marketing: What Parents Can Do.* http://www.media-awareness.ca/english/parents/marketing/dealing_marketing.cfm.

Vegetarianism and Veganism

Books

Bittman, M. *Food Matters: A Guide to Conscious Eating.* New York: Simon and Schuster, 2008.

———. *How to Cook Everything Vegetarian.* Hoboken, NJ: Wiley Publishing Inc., 2007.

Brown, E. E. *The Complete Tassajara Cookbook: Recipes, Techniques, and Reflections from the Famed Zen Kitchen.* Boston: Shambhala, 2009.

Davis B., and V. Melina. *Becoming Vegan: The Complete Guide to Adopting a Healthy Plant-Based Diet.* Summertown, TN: Book Publishing Company, 2000.

Foer, J. S. *Eating Animals.* New York: Little, Brown and Company, 2009.

Hutchins, I., and D. M. Daniels. *The Vegetarian Soul Food Cookbook.* New York: Epiphany Books, 2001.

Jacobson, M. *Six Arguments for a Greener Diet: How a More Plant-Based Diet Could Save Your Health and the Environment.* Washington, DC: Center for Science in the Public Interest, 2006.

Katzen, M. *The New Moosewood Cookbook.* Berkeley: Ten Speed Press, 2000.

———. *The Vegetable Dishes I Can't Live Without.* New York: Hyperion, 2007.

Laland, S. *Peaceful Kingdom: Random Acts of Kindness by Animals.* Berkeley: Conari Press, 1997.

Lappe, A., and B. Terry. *Grub: Ideas for an Urban Organic Kitchen.* New York: Penguin Books, 2006,

Ly, Thi Phung, A. McIvor, J. Magezis, and the Cambridge Sangha. *Vegetarian Vietnamese Cookery.* Cambridge, UK: Cambridge Sangha. Available from the Community of Interbeing Bookshop, tel: 0870 850 2615, e-mail: book shop@interbeing.org.uk, www.interbeing.org.uk/bookshop/.

Madison, D. *Vegetarian Cooking for Everyone.* New York: Broadway Books, 1997.

Medearis, A. S. *The Ethnic Vegetarian: Traditional and Modern Recipes from Africa, America, and the Caribbean.* New York: Rodale Inc., 2004.

Moskowitz, I. C., and T. H. Romero. *Veganomicon: The Ultimate Vegan Cookbook.* New York: Marlowe and Company, 2007.

Terry, B. *Vegan Soul Kitchen: Fresh, Healthy, and Creative African-American Cuisine.* Cambridge, MA: Da Capo Press, 2009.

Web Sites and Online Publications

Boston Vegan Association. *Frequently Asked Questions.* http://bostonvegan .org/index.php?option=com_content&task=category§ionid=3&id=6 0&Itemid=43.

Johns Hopkins University Bloomberg School of Public Health. *Meatless Monday Home Page.* http://www.meatlessmonday.com/.

Pew Commission on Industrial Farm Animal Production. *Putting Meat on the Table: Industrial Farm Animal Production in America.* http://www.ncifap.org/_images/PCIFAPFin.pdf.

Physicians Committee for Responsible Medicine. *Physicians Committee for Responsible Medicine Home Page.* http://www.pcrm.org/.

Steinfeld, H., P. Gerber, T. Wassenaar, V. Castel, V., M. Rosales. and C. de Haan. *Livestock's Long Shadow: Environmental Issues and Options.* Rome: Food & Agriculture Organization of the UN, 2006. http://www.fao.org/docrep/010/a0701e/a0701e00.HTM.

Union of Concerned Scientists. *Union of Concerned Scientists Home Page.* http://www.ucsusa.org/.

Vegetarian Resource Group. *The Vegetarian Resource Group Home Page.* http://www.vrg.org/.

Vegetarian Society. *The Vegetarian Society Home Page.* http://www.vegsoc.org/.

Vegetarian Society (Singapore). *The Vegetarian Society (Singapore) Home Page.* http://www.vegetarian-society.org/.

NOTES

Chapter 1: Ending Your Struggle with Weight

1. C. L. Ogden et al., Prevalence of overweight and obesity in the United States, 1999–2004, *JAMA* 295 (2006): 1549–55.
2. D. A. Kessler, *The End of Overeating: Taking Control of the Insatiable American Appetite* (New York: Rodale, 2009).
3. The Nielsen Company, NielsenWire, Ad Spending Down 11.5 Percent in First Three Quarters of 2009. Updated December 10, 2009. Accessed December 18, 2009. http://blog.nielsen.com/nielsenwire/consumer/ad-spending-in-u-s-down-11-5-percent-in-first-three-quarters-of-2009/.
4. NielsenWire, More than half the homes in U.S. have three or more TVs (2009), http://blog.nielsen.com/nielsenwire/media_entertainment/more-than-half-the-homes-in-us-have-three-or-more-tvs/.
5. Marketdata Enterprises. Press release: Diet Market Worth $58.6 Billion in U.S. Last Year, but Growth Is Flat, Due to the Recession. Tampa FL, February 16, 2009. http://www.marketdataenterprises.com/pressreleases/DietMkt2009PressRelease.pdf. Accessed November 30, 2009.
6. R. S. Padwal and S. R. Majumdar. Drug treatments for obesity: orlistat, sibutramine, and rimonabant, *Lancet* (2007); 369:71–77.11.
7. T. N. Hanh, *The Heart of the Buddha's Teaching* (Berkeley, CA: Parallax Press, 1998).
8. T. N. Hanh. *Thich Nhat Hanh 2008 Calendar.* Brush Dance, San Rafael, CA.
9. A. Anandacoomarasamy et al., The impact of obesity on the musculoskeletal system, *Int J Obes* 32 (2007): 211–22.
10. K. M. McClean et al., Obesity and the lung: 1. Epidemiology, *Thorax* 63 (2008): 649–54.
11. D. P. Guh et al., The incidence of co-morbidities related to obesity and overweight: A systematic review and meta-analysis, *BMC Public Health* 9 (2005): 88 doi: 10.1186/1471-2458-9-88.
12. World Cancer Research Fund, American Institute for Cancer Research, *Food, Nutrition, Physical Activity, and the Prevention of Cancer: A Global Perspective* (Washington, DC: AICR, 2007).

13. F. B. Hu, *Obesity Epidemiology.* (New York: Oxford University Press, 2008).
14. J. R. Loret de Mola, Obesity and its relationship to infertility in men and women. *Obstet Gynecol Clin North Am*, 36(2) (2009): 333–46, ix.
15. N. Cheung and T. Y. Wong, Obesity and eye diseases, *Survey of Ophthalmology* 52 (2007): 180–95.
16. M. A. Beydoun and Y. W. Hab, Obesity and central obesity as risk factors for incident dementia and its subtypes: A systematic review and meta-analysis, *Obesity Reviews* 9 (2008): 204–18.
17. K. F. Adams et al., Overweight, obesity, and mortality in a large prospective cohort of persons 50 to 71 years old, *N Engl J Med* 355 (2006): 763–78.
18. J. E. Manson et al., Body weight and mortality among women, *N Engl J Med* 333 (1995): 677–85.
19. R. M. Puhl and K. D. Brownell, Psychosocial origins of obesity stigma: Toward changing a powerful and pervasive bias, *Obesity Reviews* 4 (2003): 213–27. R. M. Puhl and J. D. Latner, Stigma, obesity, and the health of the nation's children. *Psychology Bulletin* 133 (2007): 557–80.
20. R. C. Whitaker et al., Predicting obesity in young adulthood from childhood and parental obesity, *N Engl J Med* 337 (1997): 869–73.
21. J. K. Lake, C. Power, and T. J. Cole, Child to adult body mass index in the 1958 British birth cohort: Associations with parental obesity, *Arch Dis Child* 77 (1997): 376–80. J. J. Reilly et al., Early life risk factors for obesity in childhood: Cohort study, *BMJ* 330 (2005): 1357.
22. T. Harder, R. Bergmann, G. Kallischnigg, and A. Plagemann, Duration of breastfeeding and risk of overweight: a meta-analysis, *Am J Epidemiol* (2005); 162: 397–403.
23. D. S. Ludwig, K. E. Peterson, and S. L. Gortmaker, Relation between consumption of sugar-sweetened drinks and childhood obesity: A prospective, observational analysis, *Lancet* 357 (2001): 505–8.
24. M. B. Schulze et al., Sugar-sweetened beverages, weight gain, and incidence of type 2 diabetes in young and middle-aged women, *JAMA* 292 (2004): 927–34.
25. L. R. Vartanian, M. B. Schwartz, and K. D. Brownell, Effects of soft drink consumption on nutrition and health: A systematic review and meta-analysis, *Am J Public Health* 97 (2007): 667–75.
26. F. B. Hu et al., Television watching and other sedentary behaviors in relation to risk of obesity and type 2 diabetes mellitus in women, *JAMA* 289 (2003): 1785–91.
27. S. R. Patel and F. B. Hu, Short sleep duration and weight gain: A systematic review, *Obesity* (Silver Spring) 16 (2008): 643–53.
28. S. R. Patel et al., Association between reduced sleep and weight gain in women, *American Journal of Epidemiology* 164 (2006): 947–54.
29. Patel and Hu, Short sleep duration.
30. Patel and Hu, Short sleep duration.

31. K. L. Knutson and E. Cauter, Associations between sleep loss and increased risk of obesity and diabetes, *Ann NY Acad Sci* 1129 (2008): 287–304.
32. K. Spiegel et al., Brief communication: Sleep curtailment in healthy young men is associated with decreased leptin levels, elevated ghrelin levels, and increased hunger and appetite, *Annals of Internal Medicine* 141 (2004): 846.
33. B. J. Rolls, The supersizing of America: Portion size and the obesity epidemic, *Nutr Today* 38 (2003): 42–53.
34. B. Wansink and S. Park, At the movies: How external cues and perceived taste impact consumption volume, *Food Quality and Preference* 12 (2001): 69–74.
35. B. Wansink and M. M. Cheney, Super bowls: Serving bowl size and food consumption, *JAMA* 293 (2005): 1727–28.
36. B. Wansink and K. van Ittersum, Portion size me: Downsizing our consumption norms, *Journal of the American Dietetic Association* 107 (2007): 1103–6.
37. A. Bandura, *Self-Efficacy: The Exercise of Control* (New York: W. H. Freeman and Company, 1995).
38. R. R. Wing and J. O. Hill, Successful weight loss maintenance, *Annu Rev Nutr* 21 (2001): 323–41 S. Phelan, et al. Are the eating and exercise habits of successful weight losers changing? *Obesity* 14 (2006): 710–716. D. A. Raynor, et al. Television viewing and long-term weight maintenance: Results from the National Weight Control Registry, *Obesity* 14 (2006): 1816–24.

Chapter 2: Are You Really Appreciating the Apple? An Apple Meditation

1. J. Mooallem, Twelve easy pieces, *New York Times*, February 12, 2006.

Chapter 3: You Are *More* Than What You Eat

The epigraph is taken from the Sutra of the Son's Flesh. The sutra is included in its entirety in appendix B. For commentary on the sutra from which the epigraph is taken, see Thich Nhat Hanh, *The Path of Emancipation* (Berkeley, CA: Parallax Press, 2000), 84–91.

1. C. R. Gail et al., Breakfast habits, nutritional status, body weight, and academic performance in children and adolescents, *Journal of the American Dietetic Association* 105 (2005): 743–60.
2. J. J. Ratey, *Spark: The Revolutionary New Science of Exercise and the Brain* (Boston: Little, Brown & Company, 2008).
3. L. D. Kubzansky, Sick at heart: The pathophysiology of negative emotions, *Cleve Clin J Med* 74, suppl. 1 (2007): S67–S72.
4. M. Nestle, *What to Eat* (New York: North Point Press, 2006).
5. USDA Economic Research Service, *Food CPI, Prices and Expenditures: Food Expenditures by Families and Individuals as a Share of Disposable Personal Income* (Washington, DC, 2008).
6. H. Steinfeld et al., *Livestock's Long Shadow: Environmental Issues and Options* (Rome: Food and Agriculture Organization of the United Nations, 2006).

7. D. Pimentel and M. Pimentel, Sustainability of meat-based and plant-based diets and the environment, *Am J Clin Nutr* 78 (2003): 660S–663S.
8. Pimentel and Pimentel, Sustainability.
9. U.S. Environmental Protection Agency. Major Crops Grown in the United States. www.epa.gov/oecaagct/ag101/cropmajor.html. Last updated Thursday, September 10, 2009. Accessed December 19, 2009.
10. R. E. Black, L. H. Allen, Z. A. Bhutta, et al. Maternal and child undernutrition: Global and regional exposures and health consequences, *Lancet* (2008); 371: 243–260.
11. Pew Commission on Industrial Farm Animal Production, *Putting Meat on the Table: Industrial Farm Animal Production in America* (Washington, DC: Pew Charitable Trust and Johns Hopkins Bloomberg School of Public Health, 2008).
12. Pew Commission, *Putting Meat on the Table.*
13. Steinfeld et al., *Livestock's Long Shadow.*
14. Steinfeld et al., *Livestock's Long Shadow.*
15. Steinfeld et al., *Livestock's Long Shadow.*
16. W. J. Craig, Health effects of vegan diets, *Am J Clin Nutr* 89 (2009): 1627S–1633S. G. E. Fraser, Vegetarian diets: What do we know of their effects on common chronic diseases? *Am J Clin Nutr* 89 (2009): 1607S–1612S.
17. World Cancer Research Fund, American Institute for Cancer Research, *Food, Nutrition, Physical Activity, and the Prevention of Cancer: A Global Perspective* (Washington, DC: AICR, 2007). R. Sinha et al., Meat intake and mortality: A prospective study of over half a million people, *Arch Intern Med* 169 (2009): 562–71.
18. ZenithOptimedia, Press release: Global ad market to accelerate in 2008 despite credit squeeze (2007).
19. A. Mathews et al., *The Marketing of Unhealthy Food to Children in Europe: A Report of Phase 1 of the Children, Obesity, and Associated Chronic Diseases Project* (Brussels: European Heart Network, 2005).
20. V. Vicennati et al., Stress-related development of obesity and cortisol in women, *Obesity* 17 (2009): 1678–83. T. C. Adam and E. S. Epel, Stress, eating and the reward system, *Physiology & Behavior* 91 (2007): 449–58.

Chapter 4: Stop and Look: The Present Moment

1. T. N. Hanh, *Transformation and Healing* (Berkeley, CA: Parallax Press, 2006), p. 89.

Chapter 5: Mindful Eating

1. Department of Nutrition, Harvard School of Public Health, *The Nutrition Source: Knowledge for Healthy Eating.* http://www.thenutritionsource.org.
2. J. S. L. de Munter et al., Whole grain, bran, and germ intake and risk of type 2 diabetes: A prospective cohort study and systematic review, *PLoS Medicine* (4) 2007: e261. S. Liu et al., Whole-grain consumption and risk of coronary

heart disease: Results from the Nurses' Health Study, *Am J Clin Nutr* 70 (1999): 412–19. P. B. Mellen, T. F. Walsh, and D. M. Herrington, Whole grain intake and cardiovascular disease: A meta-analysis, *Nutrition, Metabolism and Cardiovascular Diseases* 18 (2008): 283–90.

3. A. Schatzkin et al., Dietary fiber and whole-grain consumption in relation to colorectal cancer in the NIH-AARP Diet and Health Study, *Am J Clin Nutr* 85 (2007): 1353–60.
4. V. S. Malik and F. B. Hu, Dietary prevention of atherosclerosis: Go with whole grains, *Am J Clin Nutr* 85 (2007): 1444–45.
5. L. Djousse et al., Egg consumption and risk of type 2 diabetes in men and women, *Diabetes Care* 32 (2009) 295–300. F. B. Hu et al., A prospective study of egg consumption and risk of cardiovascular disease in men and women, *JAMA* 281 (1999); 1387–94.
6. Institute of Medicine, *Dietary Reference Intakes for Energy, Carbohydrate, Fiber, Fat, Fatty Acids, Cholesterol, Protein, and Amino Acids (Macronutrients)* (Washington, DC: The National Academies Press, 2005). W. J. Craig and A. R. Mangels, Position of the American Dietetic Association: Vegetarian diets. *J Am Diet Assoc* 109(7) (2009): 1266–82.
7. R. P. Mensink et al., Effects of dietary fatty acids and carbohydrates on the ratio of serum total to HDL cholesterol and on serum lipids and apolipoproteins: A meta-analysis of 60 controlled trials, *Am J Clin Nutr* 77 (2003): 1146–55.
8. D. Mozaffarian, A. Aro, and W. C. Willett, Health effects of trans-fatty acids: Experimental and observational evidence, *Eur J Clin Nutr* 63 (2009): S5-S21. D. Mozaffarian et al., Trans fatty acids and cardiovascular disease, *N Engl J Med* 354 (2006): 1601–13.
9. Mozaffarian, Aro, and Willett, Health effects of trans-fatty acids. D. Mozaffarian, Trans fatty acids: Effects on systemic inflammation and endothelial function, *Atherosclerosis Supplements* 7 (2006): 29–32.
10. Mozaffarian, Aro, and Willett, Health effects of trans-fatty acids.
11. E. W. T. Chong et al., Fat consumption and its association with age-related macular degeneration, *Arch Ophthalmol* 127 (2009): 674–80.
12. D. Mozaffarian, Fish and n–3 fatty acids for the prevention of fatal coronary heart disease and sudden cardiac death, *Am J Clin Nutr* 87 (2008): 1991S–1996S.
13. P. C. Calder, n-3 polyunsaturated fatty acids, inflammation, and inflammatory diseases, *Am J Clin Nutr* 83 (2006): S1505–S1519.
14. D. Mozaffarian and E. B. Rimm, Fish intake, contaminants, and human health: Evaluating the risks and the benefits, *JAMA* 296 (2006): 1885–99. GISSI Prevenzione Investigators, Dietary supplementation with n–3 polyunsaturated fatty acids and vitamin E after myocardial infarction: Results of the GISSI-Prevenzione trial, *Lancet* 354 (1999): 447–55. M. Yokoyama et al., Effects of eicosapentaenoic acid on major coronary events in hypercholesterolaemic patients

(JELIS): A randomised open-label, blinded endpoint analysis, *Lancet* 369 (2007): 1090–98.

15. A. Baylin et al., Adipose tissue Alpha-linolenic acid and nonfatal acute myocardial infarction in Costa Rica, *Circulation* 197 (2003): 1586–91. H. Campos, A. Baylin, and W. C. Willett, Alpha-linolenic acid and risk of nonfatal acute myocardial infarction, *Circulation* 118 (2008): 339–45.
16. F. M. Sacks et al., Comparison of weight-loss diets with different compositions of fat, protein, and carbohydrates, *N Engl J Med* 360 (2009): 859–73. I. Shai I et al., Weight loss with a low-carbohydrate, Mediterranean, or low-fat diet, *N Engl J Med* 359 (2008): 229–41.
17. F. B. Hu and W. C. Willett, Optimal diets for prevention of coronary heart disease, *JAMA* 288 (2002): 2569–78. M. B. Schulze and F. B. Hu, Primary prevention of diabetes: What can be done and how much can be prevented? *Annual Review of Public Health* 26 (2005): 445–67.
18. P. N. Singh, J. Sabate, and G. E. Fraser, Does low meat consumption increase life expectancy in humans? *Am J Clin Nutr* 78 (2003): 526S–532S.
19. T. J. Key et al., Mortality in British vegetarians: Results from the European Prospective Investigation into Cancer and Nutrition (EPIC-Oxford), *Am J Clin Nutr* 89 (2009): 1613S–1619S.
20. W. J. Craig, Health effects of vegan diets, *Am J Clin Nutr* (2009): 1627S–33S. G. E. Fraser, Vegetarian diets: What do we know of their effects on common chronic diseases? *Am J Clin Nutr* 89 (2009): 1607S–1612S.
21. T. J. Key et al., Cancer incidence in vegetarians: Results from the European Prospective Investigation into Cancer and Nutrition (EPIC-Oxford), *Am J Clin Nutr* 89 (2009): 1620S–1626S.
22. Craig, Health effects of vegan diets.
23. M. Jacobson, *Six Arguments for a Greener Diet: How a Plant-Based Diet Could Save Your Health and the Environment* (Washington, DC: Center for Science in the Public Interest, 2006).
24. A. Vang et al., Meats, processed meats, obesity, weight gain and occurrence of diabetes among adults: Findings from Adventist Health Studies, *Ann Nutr Metab* 52 (2008): 96–104.
25. World Cancer Research Fund, American Institute for Cancer Research, *Food, Nutrition, Physical Activity, and the Prevention of Cancer: A Global Perspective* (Washington, DC: AICR, 2007).
26. R. Z. Stolzenberg-Solomon et al., Meat and meat-mutagen intake and pancreatic cancer risk in the NIH-AARP cohort, *Cancer Epidemiol Biomarkers Prev* 16 (2007): 2664–75.
27. E. Linos et al., Red meat consumption during adolescence among premenopausal women and risk of breast cancer, *Cancer Epidemiol Biomarkers Prev* 17 (2008): 2146–51.
28. T. T. Fung et al., Dietary patterns, meat intake, and the risk of type 2 diabetes in women, *Archives of Internal Medicine* 164 (2004): 2235–40.

29. F. B. Hu et al., Prospective study of major dietary patterns and risk of coronary heart disease in men, *Am J Clin Nutr* 72 (2000): 912–21.
30. T. T. Fung et al., Prospective study of major dietary patterns and stroke risk in women, *Stroke* 35 (2004): 2014–19.
31. R. Varraso et al., Prospective study of dietary patterns and chronic obstructive pulmonary disease among U.S. men, *Thorax* 62 (2007): 786–91.
32. C. Heidemann et al., Dietary patterns and risk of mortality from cardiovascular disease, cancer, and all causes in a prospective cohort of women, *Circulation* 118 (2008): 230–37.
33. A. Mente et al., A systematic review of the evidence supporting a causal link between dietary factors and coronary heart disease, *Arch Intern Med* 169 (2009): 659–69. T. T. Fung et al., Mediterranean diet and incidence of and mortality from coronary heart disease and stroke in women, *Circulation* 119 (2009): 1093–1100. F. Sofi et al., Adherence to Mediterranean diet and health status: Meta-analysis, *BMJ* 337 (2008): 668–81. P. N. Mitrou et al., Mediterranean dietary pattern and prediction of all-cause mortality in a US population: Results from the NIH-AARP Diet and Health Study, *Arch Intern Med* 167 (2007): 2461–68.
34. World Cancer Research Fund, *Food*. L. Dauchet et al., Fruit and vegetable consumption and risk of coronary heart disease: A meta-analysis of cohort studies, *J Nutr* 136 (2006): 2588–93. F. J. He et al., Increased consumption of fruit and vegetables is related to a reduced risk of coronary heart disease: Meta-analysis of cohort studies, *Journal of Human Hypertension* 21 (2007): 717–28. A.-H. Harding et al., Plasma vitamin C level, fruit and vegetable consumption, and the risk of new-onset type 2 diabetes mellitus: The European prospective investigation of cancer—Norfolk Prospective Study, *Arch Intern Med* 168 (2008): 1493–99. L. A. Bazzano et al., Intake of fruit, vegetables, and fruit juices and risk of diabetes in women, *Diabetes Care* 31 (2008): 1311–17.
35. L. Brown et al., A prospective study of carotenoid intake and risk of cataract extraction in US men, *Am J Clin Nutr* 70 (1999): 517–24. W. G. Christen et al., Fruit and vegetable intake and the risk of cataract in women, *Am J Clin Nutr* 81 (2005): 1417–22. S. M. Moeller et al., Overall adherence to the dietary guidelines for Americans is associated with reduced prevalence of early age-related nuclear lens opacities in women, *J Nutr* 134 (2004): 1812–19. E. Cho et al., Prospective study of intake of fruits, vegetables, vitamins, and carotenoids and risk of age-related maculopathy, *Arch Ophthalmol* 122 (2004): 883–92.
36. Dauchet et al., Fruit and vegetable consumption. He et al., Increased consumption of fruit and vegetables.
37. Bazzano, Intake of Fruit.
38. J. W. Anderson et al., Carbohydrate and fiber recommendations for individuals with diabetes: A quantitative assessment and meta-analysis of the evidence, *J Am Coll Nutr* 23 (2004): 5–17. T. L. Halton et al., Low-carbohydrate-diet score and the risk of coronary heart disease in women, *N Engl J Med* 355 (2006):

1991–2002. J. W. Beulens et al., High dietary glycemic load and glycemic index increase risk of cardiovascular disease among middle-aged women: A population-based follow-up study, *J Am Coll Cardiol* 50 (2007): 14–21.

39. K. C. Maki et al., Effects of a reduced-glycemic-load diet on body weight, body composition, and cardiovascular disease risk markers in overweight and obese adults, *Am J Clin Nutr* 85 (2007): 724–34. C. B. Ebbeling et al., Effects of a low-glycemic load vs. low-fat diet in obese young adults: A randomized trial, *JAMA* 297 (2007): 2092–2102.
40. R. K. Johnson, L. J. Appel, M. Brands, et al. Dietary sugars intake and cardiovascular health: A scientific statement from the American Heart Association. *Circulation* 120 (2009): 1011–20.
41. D. E. Wallis, S. Penckofer, and G. W. Sizemore, The "sunshine deficit" and cardiovascular disease, *Circulation* 118 (2008): 1476–85. M. F. Holick, Vitamin D: A D-lightful health perspective, *Nutrition Reviews* 66 (2008): S182–S194.
42. M. F. Holick, Vitamin D deficiency, *N Engl J Med* 357 (2007): 266–81.
43. D. Feskanich, V. Singh, W. C. Willett, and G. A. Colditz. Vitamin A intake and hip fractures among postmenopausal women. *JAMA* 287 (2002): 47–54. K. Michaelsson, H. Lithell, B. Vessby, and H. Melhus. Serum retinol levels and the risk of fracture. *N Engl J Med* 348 (2003): 287–94. K. L. Penniston and S. A. Tanumihardjo. The acute and chronic toxic effects of vitamin A. *Am J Clin Nutr* 83 (2006): 191–201. V. Azais-Braesco and G. Pascal. Vitamin A in pregnancy: Requirements and safety limits. *Am J Clin Nutr* 71 (2000): 1325S–33S.
44. N. R. Cook et al., Long term effects of dietary sodium reduction on cardiovascular disease outcomes: Observational follow-up of the Trials of Hypertension Prevention (TOHP), *BMJ* 334 (2007): 885.
45. American Heart Association. News Releases. American Heart Association supports lower sodium limits for most Americans. March 26, 2009. Available at http://americanheart.mediaroom.com/index.php?s=43&item=700. Accessed December 19, 2009. Centers for Disease Control and Prevention (CDC). Application of lower sodium intake recommendations to adults—United States, 1999–2006. *MMWR Morbity and Mortality Weekly Report* 58 (2009): 281–83.
46. Center for Science in the Public Interest, Xtreme Eating Awards 2009 (2009), http://www.cspinet.org/new/200906021.html.
47. K. He et al., Association of monosodium glutamate intake with overweight in Chinese adults: The INTERMAP study, *Obesity* 16 (2008): 1875–80.
48. He et al., Association of monosodium glutamate intake.
49. W. C. Willett and P. Skerrett, *Eat, Drink, and Be Healthy: The Harvard Medical School Guide to Healthy Eating* (New York: Free Press/Simon & Schuster, 2005).
50. E. Cho et al., Dairy foods, calcium, and colorectal cancer: A pooled analysis of 10 cohort studies, *Journal of the National Cancer Institute* 96 (2004): 1015–22.
51. W. Owusu et al., Calcium intake and the incidence of forearm and hip fractures among men, *Journal of Nutrition* 127 (1997): 1782–87. D. Feskanich et al., Milk, dietary calcium, and bone fractures in women: A 12-year prospective study,

American Journal of Public Health 87 (1997): 992–97. H. A. Bischoff-Ferrari et al., Calcium intake and hip fracture risk in men and women: A meta-analysis of prospective cohort studies and randomized controlled trials, *Am J Clin Nutr* 86 (2007): 1780–90.

52. World Cancer Research Fund, *Food.* J. M. Genkinger et al., Dairy products and ovarian cancer: A pooled analysis of 12 cohort studies, *Cancer Epidemiology, Biomarkers and Prevention* 15 (2006): 364–72. E. Giovannucci et al., Risk factors for prostate cancer incidence and progression in the health professionals follow-up study, *International Journal of Cancer* 121 (2007): 1571–78. E. Giovannucci et al., Calcium and fructose intake in relation to risk of prostate cancer, *Cancer Research* 58 (1998): 442–47.
53. M. B. Schulze et al., Sugar-sweetened beverages, weight gain, and incidence of type 2 diabetes in young and middle-aged women, *JAMA* 292 (2004): 927–34. V. S. Malik, M. B. Schulze, and F. B. Hu, Intake of sugar-sweetened beverages and weight gain: A systematic review, *Am J Clin Nutr* 84 (2006): 274–88. V. S. Malik, W. C. Willett, and F. B. Hu, Sugar-sweetened beverages and BMI in children and adolescents: Reanalyses of a meta-analysis, *Am J Clin Nutr* 89 (2009): 438–39; author reply, 439–40. L. R. Vartanian, M. B. Schwartz, and K. D. Brownell, Effects of soft drink consumption on nutrition and health: A systematic review and meta-analysis, *Am J Public Health* 97 (2007): 667–75. T. T. Fung et al., Sweetened beverage consumption and risk of coronary heart disease in women, *Am J Clin Nutr* 89 (2009): 1037–42. J. R. Palmer et al., Sugar-sweetened beverages and incidence of type 2 diabetes mellitus in African American women, *Arch Intern Med* 168 (2008): 1487–92.
54. Department of Nutrition, Harvard School of Public Health, How sweet is it? (2009), http://www.hsph.harvard.edu/nutritionsource/healthy-drinks/how-sweet-is-it/.
55. R. M. van Dam et al., Coffee, caffeine, and risk of type 2 diabetes: A prospective cohort study in younger and middle-aged U.S. women, *Diabetes Care* 29 (2006): 398–403. S. Kuriyama et al., Green tea consumption and mortality due to cardiovascular disease, cancer, and all causes in Japan: The Ohsaki study, *JAMA* 296 (2006): 1255–65.
56. World Cancer Research Fund, *Food.* U.S. Department of Agriculture, *Dietary Guidelines for Americans 2005* (Washington, DC: U.S. Department of Agriculture, 2005).
57. J. Rehm et al., Global burden of disease and injury and economic cost attributable to alcohol use and alcohol-use disorders, *Lancet* 373 (2009): 2223–33.
58. I. J. Goldberg et al., AHA Science Advisory: Wine and your heart: A science advisory for healthcare professionals from the Nutrition Committee, Council on Epidemiology and Prevention, and Council on Cardiovascular Nursing of the American Heart Association, *Circulation* 103 (2001): 472–75. L. L. Koppes et al., Moderate alcohol consumption lowers the risk of type 2 diabetes: A meta-analysis of prospective observational studies, *Diabetes Care* 28 (2005): 719–25.

K. M. Conigrave et al., A prospective study of drinking patterns in relation to risk of type 2 diabetes among men, *Diabetes* 50 (2001): 2390–95. L. Djousse et al., Alcohol consumption and type 2 diabetes among older adults: The Cardiovascular Health Study, *Obesity* (Silver Spring) 15 (2007): 1758–65.

59. World Cancer Research Fund, *Food.*
60. U.S. Department of Agriculture, *Dietary Guidelines.*
61. N. W. Gilpin and G. F. Koob, Neurobiology of alcohol dependence: Focus on motivational mechanisms, *Alcohol Research & Health* 31 (2008): 185–95.
62. L. Fontana and S. Klein, Aging, adiposity, and calorie restriction, *JAMA* 297 (2007): 986–94. L. K. Heilbronn et al., Effect of 6-month calorie restriction on biomarkers of longevity, metabolic adaptation, and oxidative stress in overweight individuals: A randomized controlled trial, *JAMA* 295 (2006): 1539–48.
63. G. Eshel and P. A. Martin, Diet, energy, and global warming, *Earth Interactions* 10 (2006):1–17.
64. C. L. Weber and H. S. Matthews, Food-miles and the relative climate impacts of food choices in the United States, *Environmental Science & Technology* 42 (2008): 3508–13.
65. K. Glanz et al., Psychosocial correlates of healthful diets among male auto workers, *Cancer Epidemiol Biomarkers Prev* 7 (1998): 119–26. A. R. Shaikh et al., Psychosocial predictors of fruit and vegetable consumption in adults: A review of the literature, *Am J Prev Med* 34 (2008): 535–43. J. G. Sorensen et al., The influence of social context on changes in fruit and vegetable consumption: Results of the healthy directions studies, *American Journal of Public Health* 97 (2007): 1216–27.
66. J. F. Sallis and K. Glanz, Physical activity and food environments: Solutions to the obesity epidemic, *Milbank Quarterly* 87 (2009): 123–54.
67. Sallis and Glanz, Physical activity and food environments.
68. M. S. Townsend et al., Less-energy-dense diets of low-income women in California are associated with higher energy-adjusted diet costs, *Am J Clin Nutr* 89 (2009): 1220–26. M. Maillot et al., Low energy density and high nutritional quality are each associated with higher diet costs in French adults, *Am J Clin Nutr* 86 (2007): 690–96. A. Drewnowski and N. Darmon, The economics of obesity: Dietary energy density and energy cost, *Am J Clin Nutr* 82 (2005): 265S–73S. H. Schroder, J. Marrugat, and M. I. Covas, High monetary costs of dietary patterns associated with lower body mass index: A population-based study, *Int J Obes* 30 (2006): 1574–79.
69. A. K. Kant and B. I. Graubard, Secular trends in patterns of self-reported food consumption of adult Americans: NHANES 1971–1975 to NHANES 1999–2002, *Am J Clin Nutr* 84 (2006): 1215–23. P. S. Haines, D. K. Guilkey, and B. Popkin, Trends in breakfast consumption in US adults between 1965 and 1991, *Journal of the American Dietetic Association* 96 (1996): 464–70.
70. W. O. Song et al., Is consumption of breakfast associated with body mass index in US adults? *Journal of the American Dietetic Association* 105 (2005): 1373–82.

M. T. Timlin et al., Breakfast eating and weight change in a 5-year prospective analysis of adolescents: Project EAT (Eating Among Teens), *Pediatrics* 121 (2008): e638–e645. M. T. Timlin and M. A. Pereira, Breakfast frequency and quality in the etiology of adult obesity and chronic diseases, *Nutrition Reviews* 65 (2007): 268–81. A. A. W. A. van der Heijden et al., A prospective study of breakfast consumption and weight gain among U.S. men, *Obesity* 15 (2007): 2463–69.

71. H. R. Wyatt et al., Long-term weight loss and breakfast in subjects in the National Weight Control Registry, *Obesity* 10 (2002): 78–82.
72. Timlin and Pereira, Breakfast frequency and quality. W. A. M. Blom et al., Effect of a high-protein breakfast on the postprandial ghrelin response, *Am J Clin Nutr* 83 (2006): 211–20.
73. H. M. Seagle et al., Position of the American Dietetic Association: Weight management, *Journal of the American Dietetic Association* 109 (2009): 330–46.
74. R. Stuart and B. Davis, *Slim Chance in a Fat World* (Champaign, IL: Research Press, 1972).
75. T. A. Spiegel, Rate of intake, bites, and chews: The interpretation of lean-obese differences, *Neuroscience & Biobehavioral Reviews* 24 (2000): 229–37. C. K. Martin et al., Slower eating rate reduces the food intake of men, but not women: Implications for behavioral weight control, *Behaviour Research and Therapy* 45 (2007): 2349–59. R. G. Laessle, S. Lehrke, and S. Dückers, Laboratory eating behavior in obesity, *Appetite* 49 (2007): 399–404.
76. A. M. Andrade, G. W. Greene, and K. J. Melanson, Eating slowly led to decreases in energy intake within meals in healthy women, *Journal of the American Dietetic Association* 108 (2008): 1186–91.
77. K. Maruyama et al., The joint impact on being overweight of self reported behaviours of eating quickly and eating until full: Cross sectional survey. *BMJ* 337 (2008): a2002. R. Otsuka et al., Eating fast leads to obesity: Findings based on self-administered questionnaires among middle-aged Japanese men and women, *Journal of Epidemiology* 16 (2006): 117–24. S. Sasaki et al., Self-reported rate of eating correlates with body mass index in 18-year-old Japanese women, *Int J Obes Relat Metab Disord* 27 (2003): 1405–10.
78. B. Wansink, Environmental factors that increase the food intake and consumption volume of unknowing consumers, *Annu Rev Nutr* 24 (2004): 455–79. B. Wansink and S. Park, At the movies: How external cues and perceived taste impact consumption volume, *Food Quality and Preference* 12 (2000): 69–74. B. Wansink, K. van Ittersum, and J. E. Painter, Ice cream illusions: Bowls, spoons, and self-served portion sizes, *American Journal of Preventive Medicine* 31 (2006): 240–43.
79. L. R. Young and M. Nestle, The contribution of expanding portion sizes to the US obesity epidemic, *American Journal of Public Health* 92 (2002): 246–49. S. J. Nielsen and B. M. Popkin, Patterns and trends in food portion sizes, 1977–1998, *JAMA* 289 (2003): 450–53. B. Wansink and C. R. Payne, The joy of cooking too much: 70 years of calorie increases in classic recipes, *Ann Intern Med* 150 (2009): 291–92.

80. B. Wansink and K. van Ittersum, Portion size me: Downsizing our consumption norms, *Journal of the American Dietetic Association* 107 (2007): 1103–06.
81. Wansink, Environmental factors.
82. Wansink, Environmental factors.
83. P. Chandon and B. Wansink, The biasing health halos of fast food restaurant health claims: Lower calorie estimates and higher side-dish consumption intentions, *Journal of Consumer Research* 34 (2007): 301–14.
84. A. Stunkard, K. Allison, and J. Lundgren, Issues for DSM-V: Night eating syndrome, *Am J Psychiatry* 165 (2008): 424. K. C. Allison et al., Proposed diagnostic criteria for night eating syndrome. *Int J Eat Disord* (2009). E-publication ahead of print. DOI 10.1002/eat.20693.
85. Stunkard, Allison, and Lundgren, Issues for DSM-V.
86. M. E. Gluck, A. Geliebter, and T. Satov, Night eating syndrome is associated with depression, low self-esteem, reduced daytime hunger, and less weight loss in obese outpatients, *Obes Res* 9 (2001): 264–67.
87. Wansink, Environmental factors.
88. L. A. Pawlow, P. M. O'Neil, and R. J. Malcolm, Night eating syndrome: Effects of brief relaxation training on stress, mood, hunger, and eating patterns, *Int J Obes Relat Metab Disord* 27 (2003): 970–78.
89. U.S. Department Labor, U.S. Bureau of Labor Statistics, *Consumer Expenditures in 2007* (Washington, DC, 2009).
90. J. F. Guthrie, B. H. Lin, and E. Frazao, Role of food prepared away from home in the American diet, 1977–78 versus 1994–96: Changes and consequences, *J Nutr Educ Behav* 34 (2002): 140–50.
91. Wansink, Environmental factors.
92. N. Welch et al., Is the perception of time pressure a barrier to healthy eating and physical activity among women? *Public Health Nutrition* 12 (2009): 888–95. N. I. Larson et al., Food preparation by young adults is associated with better diet quality, *Journal of the American Dietetic Association* 106 (2006): 2001–7. J. Jabs and C. M. Devine, Time scarcity and food choices: An overview, *Appetite* 47 (2006): 196–204.
93. A. A. Gorin et al., Promoting long-term weight control: Does dieting consistency matter? *Int J Obes Relat Metab Disord* 28 (2003): 278–81.
94. M. L. Butryn et al., Consistent self-monitoring of weight: A key component of successful weight loss maintenance. *Obesity* (Silver Spring) 15(12) (2007): 3091–6. K. N. Boutelle et al., How can obese weight controllers minimize weight gain during the high risk holiday season? By self-monitoring very consistently, *Health Psychology* 18 (1999): 364–68. J. F. Hollis et al., Weight loss during the intensive intervention phase of the Weight-Loss Maintenance Trial, *American Journal of Preventive Medicine* 35 (2008): 118–26.
95. L. Zepeda and D. Deal, Think before you eat: Photographic food diaries as intervention tools to change dietary decision making and attitudes, *International Journal of Consumer Studies* 32 (2008): 692–98.

96. D. L. Helsel, J. M. Jakicic, and A. D. Otto, Comparison of techniques for self-monitoring eating and exercise behaviors on weight loss in a correspondence-based intervention, *Journal of the American Dietetic Association* 107 (2007): 1807–10.
97. M. F. Dallman, Stress-induced obesity and the emotional nervous system. *Trends Endocrinol Metab,* 2009, doi:10.1016/j.tem.2009.10.004. K. Elfhag and S. Rossner, Who succeeds in maintaining weight loss? A conceptual review of factors associated with weight-loss maintenance and weight regain. *Obes Rev* 6(1) (2005): 67–85. R. M. Masheb and C. M. Grilo, Emotional overeating and its associations with eating disorder psychopathology among overweight patients with binge eating disorder. *Int J Eat Disord* 39(2) (2006): 141–6.
98. M. Faith, D. B. Allison, and A. Gelleibter, Emotional eating and obesity: Theoretical considerations and practical recommendations, in *Overweight and Weight Management: The Health Professional's Guide to Understanding and Practice,* ed. S. Dalton (Sudbury, MA: Jones and Bartlett, 1997), 455. K. Elfhag and S. Rossner, Who succeeds in maintaining weight loss?: 67–85.

Chapter 6: Mindful Moving

1. Centers for Disease Control and Prevention, 2001–2007 state physical activity statistics (2007), http://apps.nccd.cdc.gov/PASurveillance/StateSumV.asp.
2. U.S. Department of Health and Human Services, *2008 Physical Activity Guidelines for Americans: Be Active, Healthy, and Happy!* (Washington, DC: U.S. Department of Health and Human Services, 2008).
3. K. H. Cooper and T. C. Cooper, *Start Strong, Finish Strong: Prescriptions for a Lifetime of Great Health* (New York: Penguin Group, 2007).
4. J. J. Ratey, *Spark: The Revolutionary New Science of Exercise and the Brain* (Boston: Little, Brown & Company, 2008).
5. S. Begley, *Train Your Mind, Change Your Brain* (New York: Ballantine Books, 2008).
6. D. J. Siegel, *The Mindful Brain* (New York: Norton, 2007).
7. K. Shaw et al., Exercise for overweight or obesity, Cochrane Database of Systematic Reviews (2006).
8. Cooper and Cooper, *Start Strong, Finish Strong.*
9. Department of Health and Human Services, *2008 Physical Activity Guidelines.*
10. Department of Health and Human Services, *2008 Physical Activity Guidelines.*
11. American College of Sports Medicine, *ACSM's Guidelines for Exercise Testing and Prescription* (Philadelphia: Lippincott Williams & Wilkins, 2006).
12. T. Hanh and W. Vriezen, *Mindful Movements: Ten Exercises for Well-Being* (Berkeley, CA: Parallax Press, 2008).
13. American College of Sports Medicine, *ACSM's Guidelines.*
14. J. E. Donnelly et al., The effects of 18 months of intermittent vs. continuous exercise on aerobic capacity, body weight and composition, and metabolic fitness in previously sedentary, moderately obese females, *Int J Obes Rela Metab*

Disord 24 (2000): 566. J. M. Jakicic et al., Effects of intermittent exercise and use of home exercise equipment on adherence, weight loss, and fitness in overweight women: A randomized trial, *JAMA* 282 (1999): 1554–60.

15. Donnelly et al., Effects of 18 months. Jakicic et al., Effects of intermittent exercise.
16. Shaw et al., Exercise for overweight or obesity.
17. W. C. Miller, D. M. Koceja, and E. J. Hamilton, A meta-analysis of the past 25 years of weight loss research using diet, exercise or diet plus exercise intervention, *Int J Obes Rela Metab Disord* 21 (1997): 941–47. J. M. Jakicic et al., American College of Sports Medicine position stand: Appropriate intervention strategies for weight loss and prevention of weight regain for adults, *Medicine & Science in Sports & Exercise* 33 (2001): 2145–56.
18. Jakicic et al., American College of Sports Medicine position stand. J. E. Donnelly et al., American College of Sports Medicine Position Stand: Appropriate physical activity intervention strategies for weight loss and prevention of weight regain for adults, *Medicine & Science in Sports & Exercise* 41 (2009): 459–71.
19. Jakicic et al., American College of Sports Medicine position stand. Donnelly et al., Effects of 18 months.
20. V. A. Catenacci et al., Physical activity patterns in the National Weight Control Registry, *Obesity* (Silver Spring) 16 (2008): 153–61.
21. The Nielsen Company, Nielsen Wire, Average TV Viewing for 2008–09 TV Season at All-Time High. http://blog.nielsen.com/nielsenwire/media_entertainment/average-tv-viewing-for-2008-09-tv-season-at-all-time-high/. Updated November 10, 2009; accessed November 30, 2009. The Nielsen Company, Historical daily viewing activity among households & persons 2+. Available at http://blog.nielsen.com/nielsenwire/wp-content/uploads/2009/11/historicalviewing.pdf. Updated November 10, 2009; accessed November 30, 2009.
22. L. A. Tucker and M. Bagwell, Television viewing and obesity in adult females, *Am J Public Health* 81 (1991): 908–11. L. A. Tucker and G. M. Friedman, Television viewing and obesity in adult males, *Am J Public Health* 79 (1989): 516–18.
23. F. B. Hu et al., Television watching and other sedentary behaviors in relation to risk of obesity and type 2 diabetes mellitus in women, *JAMA* 289 (2003): 1785–91.
24. Hu et al., Television watching. F. Kronenberg et al., Influence of leisure time physical activity and television watching on atherosclerosis risk factors in the NHLBI Family Heart Study, *Atherosclerosis* 153 (2000): 433–43. G. N. Healy et al., Television time and continuous metabolic risk in physically active adults, *Medicine & Science in Sports & Exercise* 40 (2008): 639–45.
25. E. M. Blass et al., On the road to obesity: Television viewing increases intake of high-density foods, *Physiol Behav* 88 (2006): 597–604.
26. M. Scully, H. Dixon, and M. Wakefield, Association between commercial television exposure and fast-food consumption among adults, *Public Health Nutr* 12 (2009): 105–10.

27. S. A. Bowman, Television-viewing characteristics of adults: Correlations to eating practices and overweight and health status, *Prev Chronic Dis* 3 (2006): A38.
28. Centers for Disease Control and Prevention, 2001–2007 State physical activity statistics.
29. F. F. Reichert et al., The role of perceived personal barriers to engagement in leisure-time physical activity, *American Journal of Public Health* 97 (2007): 515–19. W. C. Stutts, Physical activity determinants in adults: Perceived benefits, barriers, and self-efficacy, *AAOHN Journal* 50 (2002): 499–507.
30. N. E. Sherwood and R. W. Jeffery, The behavioral determinants of exercise: Implications for physical activity interventions, *Annual Review of Nutrition* 20 (2000): 21–44. D. M. Williams, E. S. Anderson, and R. A. Winett, A review of the outcome expectancy construct in physical activity research, *Annals of Behavioral Medicine* 29 (2005): 70–79.
31. Reichert et al., Role of perceived personal barriers. M. Stafford et al., Pathways to obesity: Identifying local, modifiable determinants of physical activity and diet, *Social Science & Medicine* 65 (2007): 1882–97. S. A. French, M. Story, and R. W. Jeffery, Environmental influences on eating and physical activity, *Ann Rev Public Health* 22 (2001): 309–35. J. F. Sallis and K. Glanz, Physical activity and food environments: Solutions to the obesity epidemic, *Milbank Quarterly* 87 (2009): 123–54.
32. Jakicic et al., Effects of intermittent exercise.
33. Department of Health and Human Services, *2008 Physical Activity Guidelines for Americans.*
34. Hanh and Vriezen, *Mindful Movements.*
35. L. A. Levine et al., Interindividual variation in posture allocation: Possible role in human obesity, *Science* 307 (2005): 584–86.
36. T. Hanh, *The World We Have* (Berkeley, CA: Parallax Press, 2008).

Chapter 7: Mindful Living Plan

1. M. Barry and J. M. Hughes, Talking dirty: The politics of clean water and sanitation, *N Engl J Med* 359 (2008): 784–87.
2. R. Tiernan, Best option is to clean and re-use waste, *Financial Times*, March 22, 2007, p. 13.
3. R. R. Wing and J. O. Hill, Successful weight loss maintenance, *Annual Review of Nutrition* 21 (2001): 323–41. J. A. Linde et al., Self-weighing in weight gain prevention and weight loss trials, *Annals of Behavioral Medicine* 30 (2005): 210–16.
4. K. N. Boutelle et al., How can obese weight controllers minimize weight gain during the high risk holiday season? By self-monitoring very consistently, *Health Psychology* 18 (1999): 364–68. M. L. Butryn et al., Consistent self-monitoring of weight: A key component of successful weight loss maintenance, *Obesity* (Silver Spring) 15 (12) (2007): 3091–6. J. F. Hollis et al., Weight loss during the intensive intervention phase of the Weight-Loss Maintenance Trial, *American Journal of Preventive Medicine* 35 (2008): 118–26.

5. F. M. Sacks et al., Comparison of weight-loss diets with different compositions of fat, protein, and carbohydrates, *N Engl J Med* 360 (2009): 859–73.
6. K. D. Brownell et al., The effect of couples training and partner co-operativeness in the behavioral treatment of obesity, *Behav Res Ther* 16 (1978): 323–33.
7. A. A. Gorin et al., Weight loss treatment influences untreated spouses and the home environment: Evidence of a ripple effect. *Int J Obes* (London) 32 (2008): 1678–84.
8. N. A. Christakis and J. H. Fowler, The spread of obesity in a large social network over 32 years, *N Engl J Med* 357 (2007): 370–79.
9. C. Thompson, Are your friends making you fat? *New York Times,* September 13, 2009, p. MM28.
10. WGBH Educational Foundation and Harvard Medical School Division of Sleep Medicine, *Healthy Sleep: Understanding the third of our lives we so often take for granted,* http://healthysleep.med.harvard.edu/healthy/.
11. M. G. Berman, J. Jonides, and S. Kaplan, The cognitive benefits of interacting with nature, *Psychol Sci* 19 (2008): 1207–12. R. Kaplan, S. Kaplan, and R. Ryan, *With People in Mind: Design and Management of Everyday Nature* (Washington, DC: Island Press, 1998).
12. M. E. Larimer, R. S. Palmer, and G. A. Marlatt, Relapse prevention: An overview of Marlatt's cognitive-behavioral model, *Alcohol Res Health* 23 (1999): 151–60.
13. G. A. Parks and G. A. Marlatt, Relapse prevention therapy: A cognitive-behavioral approach, *National Psychologist* vol. 9 (2000): 22.
14. G. A. Parks and G. A. Marlatt, Relapse prevention.

Chapter 8: A Mindful World

1. Reprinted by permission. J. L. Levye. We are all connected. WickedLocal Sharon with News from the Sharon Advocate. June 8, 2007. Available at http://www.wickedlocal.com/sharon/news/lifestyle/columnists/x870110339. Accessed on November 30, 2009.
2. T. N. Hanh, *Thich Nhat Hanh 2008 Calendar,* Brush Dance, San Rafael, CA.
3. L. K. Khan et al., Recommended community strategies and measurements to prevent obesity in the United States, *MMWR Recommendations and Reports* 58 (2009): 1–26.
4. B. M. Popkin, *The World Is Fat* (New York: Penguin Group, 2009).
5. L. Davies, 25 years of saving lives, *Driven,* Fall (2005), 8–17. Available at http://www.madd.org/getattachment/48e81e1b-df43-4f31-b9a1-d94d5b940e62/MADD-25-Years-of-Saving-Lives.aspx . Accessed on November 30, 2009.
6. Department of Nutrition, Harvard School of Public Health, Shining the spotlight on trans fats (2007), http://www.hsph.harvard.edu/nutritionsource/nutrition-news/transfats/.
7. Center for Science in the Public Interest, Trans fat: On the way out! (2009), http://www.cspinet.org/transfat/.

8. D. S. Ludwig, K. E. Peterson, and S. L. Gortmaker, Relation between consumption of sugar-sweetened drinks and childhood obesity: A prospective, observational analysis, *Lancet* 357 (2001): 505–8.
9. Alliance for a Healthier Generation, Clinton Foundation, and American Heart Association, Alliance for a Healthier Generation and industry leaders set healthy school beverage guidelines for U.S. schools (2006), http://www.americanheart.org/presenter.jhtml?identifier=3039339.
10. M. B. Schulze et al., Sugar-sweetened beverages, weight gain, and incidence of type 2 diabetes in young and middle-aged women, *JAMA* 292 (2004): 927–34. T. T. Fung et al., Sweetened beverage consumption and risk of coronary heart disease in women, *Am J Clin Nutr* 89 (2009): 1037–42. J. R. Palmer et al., Sugar-sweetened beverages and incidence of type 2 diabetes mellitus in African American women, *Archives of Internal Medicine* 168 (2008): 1487–92. V. S. Malik, M. B. Schulze, and F. B. Hu, Intake of sugar-sweetened beverages and weight gain: A systematic review, *Am J Clin Nutr* 84 (2006): 274–88. L. R. Vartanian, M. B. Schwartz, and K. D. Brownell, Effects of soft drink consumption on nutrition and health: A systematic review and meta-analysis, *American Journal of Public Health* 97 (2007): 667–75.
11. S. L. Mercer et al., Drawing possible lessons for obesity prevention and control from the tobacco control experience, in *Obesity Prevention & Public Health*, ed. D. Crawford and R. W. Jeffery (New York: Oxford Univ. Press, 2005), 231–63.
12. FXB International, About FXB: History. Accessed on November 29, 2009, http://www.fxb.org/AboutFXB/history.html.
13. Department of Nutrition, Harvard School of Public Health, Time to focus on healthier drinks (2009), http://www.hsph.harvard.edu/nutritionsource/healthy-drinks/focus/.
14. L. K. Khan et al., Recommended community strategies: 1–26.
15. T. N. Hanh, *Transformation at the Base* (Berkeley, CA: Parallax Press, 2001).

INDEX